Hector F.E. Jungerson

On the Appendices Genitales in the Greenland Shark,

Somniosus microcephalus (Bl. Schn.), and Other Selachians

Hector F.E. Jungerson

On the Appendices Genitales in the Greenland Shark,
Somniosus microcephalus (Bl. Schn.), and Other Selachians

ISBN/EAN: 9783337318154

Printed in Europe, USA, Canada, Australia, Japan

Cover: Foto ©berggeist007 / pixelio.de

More available books at **www.hansebooks.com**

THE DANISH INGOLF-EXPEDITION.

VOLUME II.

2.

ON THE APPENDICES GENITALES IN THE GREENLAND SHARK, *SOMNIOSUS MICROCEPHALUS* (BL. SCHN.), AND OTHER SELACHIANS.

BY

HECTOR F. E. JUNGERSEN.

WITH 6 PLATES AND 28 FIGURES IN THE TEXT.

— . —

TRANSLATED BY TORBEN LUNDBECK.

➡➤⧓⧔⬅

COPENHAGEN.
BIANCO LUNO (F. DREYER), PRINTER TO THE COURT.
1899.

CONTENTS.

On the Appendices Genitales in the Greenland Shark, Somniosus Microcephalus (Bl. Schn.), and other Selachians.

On the Appendices Genitales (Claspers) in the Greenland Shark, Somniosus microcephalus (Bl. Schn.), and other Selachians.

By

Hector F. E. Jungersen.

The following treatise has its origin from the circumstance that during the stay at Iceland of the cruiser Ingolf I endeavoured to gather informations as to several facts concerning the Greenland Shark, not yet elucidated. I succeeded only in throwing light upon a single one of these obscure facts by gathering a suitable material. At the subsequent examination of this material I soon perceived that the appendices genitales or claspers of the Selachians generally had hitherto been very imperfectly examined although these organs on account of their conspicuous   sometimes almost colossal dimensions have from time immemorial been known as characteristic for the males of cartilaginous fishes. Of their functions only little is known with certainty, and on this point I am not able to bring new facts of any importance; but though the function must be supposed to be the same in all Selachians, a rich variation is found in their structure, especially in the skeleton, the structure being different from genus to genus or even from species to species. That, however, through all this variation a common type may be shown to exist, also with respect to the skeleton and the muscles, has not hitherto been seen, but will, I hope, with sufficient clearness be shown by the following treatise. As a consequence of the way, in which the work has come into existence, I have divided it into two parts, of which one deals with the Greenland Shark only, while the other treats of other Plagiostomes and Holocephales.

I.

The Appendages of the Ventrals in the Greenland Shark.

The words with which Gunnerus[1] commences his treatise of the Greenland Shark: This fish of the Haackind deserves to be somewhat better known to the learned than hitherto it has been may be said to some extent to be in force to this day, our knowledge of this species of sharks being still rather defective, although it is not only very frequently found in the northern seas, but is also in several places the object of a large and regular fishery, as in our northern dependencies, especially off the coast of Iceland. It is so far less extraordinary, that many things with regard to

[1] Om Haa-Skierdingen. Det Throndhiemske Selskabs Skrifter. 2, 1763, p. 330.

its biological conditions are unknown, as the same thing may be said of many common species of fishes on our own coasts; but it seems more remarkable that we do not even know for certain whether the Greenland Shark is viviparous or oviparous, and that several features of the anatomical structure of the animal are unknown or only deficiently known. Although this species of Sharks is rather frequently found on the more populated European coasts — also on ours   and more than once has come into the hands of naturalists, even anatomists, we are thus far from being perfectly acquainted with the structure of its urinary and reproductive organs.

The facts which have in later years been brought forth as to the latter — and upon the whole concerning the viscera of the Greenland Shark -- are due to Sir William Turner, who has examined several specimens from British coasts and has given his results in The Journal of Anatomy and Physiology [1].

As to the female the first of these communications (1) showed the surprising result that oviducts were wanting. Consequently the Greenland Shark would necessarily be oviparous, and the ova, detached from the ovary, would presumably leave the abdominal cavity through the abdominal pores to be impregnated outside the mother. That the ovaries were immature in both the examined animals of a respective length of 11 ft. 8 inches and 8½ ft. is however evident from the description. Later (3) the first statement is corrected: oviducts [2] are found, opening as usual in the Sharks with wide, funnelshaped, closely united mouths before the liver, and running along the lower side of the kidneys to the cloaca; in the examined specimen of 7 feet length they were about as thick as a goosequill; the ovaries were quite immature. Still later (4) these parts are described in a somewhat more developed state in a Greenland Shark 11 ft. 6 in. long; the diameter of the oviduct was only ³⁄₈ inch (about 1 ctm.); the ovaries were quite immature. In none of these communications is shown, whether any shell gland , any indication of an uterus, indications of folds of the mucous membrane or the like were found. To judge from the fact of these structures not being mentioned, that nothing of the kind is found, I do not think justifiable; a shell-gland for inst. is generally always found in Sharks, whether they be oviparous or viviparous; more probably these structures on account of the immature state of the animals have not been prominent, and therefore have not been noticed. For that all the females examined by Sir W. Turner have been immature and young animals admits, I think, of no doubt. The fact is that we know to a certainty that the mature ovarial eggs are about as large as goose-eggs, but the largest mentioned by Sir W. Turner were only of the size of shot or at most of small bullets, and we know that the Greenland Shark grows to a still more considerable size than 11 ft. 8 in.; therefore if the oviducts showed so small a size and besides (presumably) so simple a shape, it is only, what might be expected in younger individuals [3], and I see no reason at all to

[1] 1) A Contribution to the Visceral Anatomy of the Greenland Shark (Laemargus borealis). l. c. 7, 1873, p. 233. 2) Additional observations on the Anatomy of the Greenl. Shark. l. c. 8, 1874, p. 285. 3) Note on the Oviducts of the Greenl. Shark. l. c. 12, 1878, p. 604. 4) Additional Note on the Oviducts etc. l. c. 19, 1885, p. 221.

[a] The oviducts had already been seen in 1847 by Kneeland (Boston Journ. Nat. Hist. 5, p. 479, 485) in a specimen of the length of 7 ft. 5 in.; the ovaries were immature. The first statement by Sir W. Turner has been repeated by Fürbringer: Zur vergl. Anat. u. Entwickelungsgesch. der Excretionsorgane der Vertebraten (Morphol. Jahrb. 4, 1878) p. 53, 85; it is found as late as in Guido Schneider: Ueber die Entw. der Genitalcanäle bei Cobitis taena L. und Phoxinus laevis Ag. (Mém. Ac. Imp. d. Sc. de St. Pétersbourg [8] T. 2, 1895) p. 9.

5) Comp. Joh. Müller: Untersuchungen über die Eingeweide der Fische, Schluss der vergleichende Anatomie der Myxinoiden (Abhdl. K. Ac. Wiss. Berlin 1843 [1845]), p. 133, 134.

suppose, as Sir W. Turner[1]) does, another mode of bringing forth the ova in this than in other Sharks; the ova certainly all get into the oviduct, and are impregnated there: whether they later are laid or develop into embryos in the uterus must for the present be left undecided[2]).

On the internal reproductive organs of the male only one communication (2) has been given, concerning a specimen of the length of 6 ft. 1 in. The testes were immature; neither by the direct examination of them and their mesorchium nor by injection from the renal duct was Sir W. Turner able to detect any duct for the sperm, and from that he infers that distinct sexual ducts are also wanting in the male, and that the sperm is evacuated into the abdominal cavity, thus quite corresponding to the case of the females, as it the previous year had been understood with regard to those; but while the statement has been corrected by T. himself with regard to the latter, nothing has as yet come to light concerning the male. I think, however, that the supposition is allowable, that T.'s inference is premature also with regard to the male; it is likely that vasa efferentia in this young, immature specimen (which T. himself declares to be of immature growth) were either not formed at all or at all events not in a directly visible way[3]). It must appear quite natural that also the external male genitals were quite undeveloped in this specimen; the copulatory appendages were only of a length of 1⅜ inch, and were far from reaching the end of the fin-membrane (see the fig. l. c. p. 287). But these copulatory appendages seem always to have shown a quite similar undeve-

[1] Sir W. Turner evidently has not been able quite to dismiss his original conception of the evacuation of the ova through the abdominal pores (to which for the rest every parallel would be wanting, as the Cyclostomes have no abdominal pores); even in his latest communication (4, 1885 p. 222) T. says: But, as it is very doubtful if the entire surface of each ovary could be embraced by the spathe-like canal (i. e. the mouth of the oviduct), a proportion of the ova would probably be shed into the peritoneal cavity, and be evacuated through the abdominal pores .

[2] Professor Lütken in: Smaa Bidrag til Selachiernes Naturhistorie. 2. Om Havkalens Forplantning (Vid. Medd. Naturh. Foren. i Kbhvn. 1879–80; p. 56) has tried to make it probable that the Greenland Shark should be oviparous, and moreover have soft, shell-less eggs. which is known in no other plagiostome. Among the reasons that might give some countenance to this notion Sir W. Turner's anatomical results are quoted. It is quite evident that if T.'s first communication of the want of oviducts had been correct, a deposition of the eggs, and an impregnation of them outside of the body of the female would have been as good as proved; but the later informations from the same author are in my opinion of such a nature, that they can be used as proofs neither (for nor against a deposition of the eggs, but might — connected with my demonstration in the following, that the male Greenland Shark has fully developed copulatory organs — be used as proofs of the eggs, as generally in Sharks, being impregnated in the oviduct. The other reasons for a deposition of the eggs, quoted by Professor L., viz. the negative one that we have never hitherto got any foetus of the Greenland Shark, and the more positive accounts from several laymen of numerous large eggs, but always in the females, cannot, I think, prove anything either in one or the other direction. Against the first of these reasons may be quoted the equally negative circumstance that we have never found eggs of the Greenland Shark outside the animal neither, and against the second that the large eggs are evidently ovarial eggs still coherent by the thin, distended ovarial stroma; for all informations — also those I have got personally from an Icelandic Shark-fisher — state that the large eggs, which are only seen by the flensing, always cohere by thin membranes or the like; but large and soft ovarial eggs, as is well known, are not only found in oviparous, but as well in viviparous Sharks and Rays. As however the only earlier authors, who state anything at all about the propagation, declare quite positively, that the Greenland Shark is viviparous, viz. besides Otto Fabricius and Faber, who are both cited by Professor Lütken, also David Crantz, who says in his Historie von Grönland , 2. Aufl. 1770 p. 138: Er bringt gemeiniglich 4 Junge zugleich zur Welt (from this work the statement is adopted by Couch, from whom Günther probably has his remark: It is stated to be viviparous, and to produce about four young at a birth (Introd. to the study of Fishes, 1880 p. 333) — and as moreover the very nearest relative of the Greenland Shark, the Somniosus rostratus of the Mediterranean, is known quite certainly to be viviparous, as also the somewhat more distant relatives, the Scymnus-species and the other Spinacidæ, I, to be sure, think it most probable — I feel tempted to use a stronger expression — that also the Greenland Shark, the other Somniosus-species, must be viviparous.

[3] According to Semper: Das Urogenitalsystem der Plagiostomen etc. (Arb. Zool. Zool. Inst. Würzburg, 2, 1875 vasa efferentia are in several Sharks already formed in the embryo; but I think it is doubtful whether they can be recognized here without the assistance of the microscope, and it does not appear that Sir W. Turner has used a microscopical examination; but he says that the mesorchium was so transparent that he must have seen a duct, if there had been one. The part of the testis itself, which T. especially examined to trace a possible duct in it, can scarcely contain such a one, as it is evidently the Vorkeimfalte of Semper, i. e. the part where the new ampullæ are formed.

loped condition in the other (and it turns out to be very few) male specimens, mentioned as examined by naturalists.

In this circumstance, in connection with the interpretations by Sir W. Turner of the genital apparatus in both sexes, is most likely to be sought the reason of the idea that the Greenland Shark only should be possessed of rudimentary copulatory appendages. This supposition has been set forth by Professor Lütken in the communication on the propagation of the Greenland Shark, cited on p. 3 note 2. In this paper Sir W. Turner's description of the reproductory organs both of the male and female is reported with the following remark: Of what use the copulatory members of the male were was not evident; but perhaps these organs are in this species of Sharks rudimentary structures without any importance? At all events I know no descriptions giving them a size like that found in the Spiny Dog-fish or the Basking Shark». I must confirm the latter sentence myself. It was to be expected beforehand that, if the male of this species had really copulatory appendages of proportions relatively as those of other species, so prominent formations would scarcely have escaped the notice, but would probably have been mentioned by one or more of the many earlier authors, who have written of the North and the Northern nature, in which writings the Greenland Shark and the catching of it bear a part, and of whom more, I suppose, have had the opportunity of knowing the animal by autopsy.

However, I have in vain sought in authors as: Egede, Cranz, O. Fabricius, Scoresby, Eggert Olafsen, Mohr, Olaus Olavius, Faber, Pontoppidan, Strom, Leem, Rosted, Landt, and others; I find nothing concerning this point. Only Gunnerus[1] mentions these organs, which we have reason to take to be the external characteristics of the male , but in undeveloped condition. Gunnerus had 3 male specimens, the largest not exceeding 5 ells (Danish) in length, and the smallest being $2\frac{1}{2}$ ell; the figure shows the appendages quite small, shorter than the fin-membrane; besides it is evident from his description, that he himself justly thinks his specimens to be young animals.

Later authors too do not mention appendages in more developed condition; they are on the whole (as far as I know) only mentioned by Yarrell and by Malm. Yarrell[2] says of a specimen described by Valenciennes[3]: The fish was a male; the ventral fins and sexual appendages or claspers very small . Valenciennes himself, however, says nothing of the sex, and does not at all mention the appendages; he only says that the ventrals are small, so that possibly the cited remark of Yarrell has it origin from a misreading. Malm[4] mentions two males, which he correctly declares to be young, respectively of a length of 1850^{mm} and 1880^{mm}; the length of the hjelpgenitalia» was in both 25^{mm}; they did not reach the end of the ventral fin. Only in one place I have found a statement suggesting, that the authors in question have had the opportunity of seeing the appendages of the Greenland Shark in a more developed state, viz. in Müller and Henle[5]. They divide the genus *Scymnus* in two subgenera: 1) *Scymnus* (to which *Sc. lichia* and *S. brasiliensis*), characterized among

[1] L. c. p. 330 seq., pl. X, fig. 1, Lit. a. Pl. XI, fig. 1, Litt. a, a.
[2] History of British Fishes, 3d ed., 2, p. 527.
[3] Nouv. Ann. du Muséum. 1, p. 455, pl. 20.
[4] Göteborgs och Bohusläns Fauna. 1877. p. 627. 629.
[5] Systematische Beschreibung der Plagiostomen. 1841. p. 91, 93.

other things by: Die männlichen Anhänge ohne Stachel , and 2) *Læmargus* (to which *L. borealis* [the Greenland Shark], *L. Labordii* and *L. rostratus*) in which: die Männchen haben einen Stachel an den Anhängen . But whence have M. & H. this latter information? The work itself tells nothing about it, and in none of the works cited is found anything about a spine on the appendage in *S. borealis* (and no more in the other species).

After Professor Lütken having given the cited communication about the propagation of the Greenland Shark the Museum of Copenhagen has got a male specimen of a length of 9 ft. (2835^{mm}), whose ventrals are preserved in the collection; also in this specimen the copulatory appendages are very small as hereafter mentioned, and so far they might serve as a corroboration of the advanced conjecture, that in this Shark these organs should be rudimentary and functionless.

As I, however, had some doubts of the correctness of this supposition — as also of the other that the Greenland Shark should be oviparous -- I endeavoured during the last cruise of the Ingolf to get fins of the Greenland Shark for examination, and as far as possible to procure reliable informations of this Shark in all respects. During a stay in the close of June 1896 in Dyrefjord, where a manufactory for train-oil of the Greenland Shark is found, I took the opportunity of communicating with a fisher of Greenland Sharks, whom I for some time questioned by means of an interpreter. The conversation was rather difficult, as the man was somewhat embarassed, only answered to my questions, and would not speak himself or give his own opinion. However I got the information that the fishermen know very well to distinguish between male and female, that eggs (i. e. the large ovarial eggs) are only found in large specimens, and that the males are smaller than the females; he had however never seen a Greenland Shark smaller than about 3 ells (Danish)[1]. I drew a sketch of the ventrals for him, and asked, if he had seen the appendages on the ventrals, which he affirmed; then I promised him a reward, if he would obtain for me as many pairs of ventrals as possible, and with as large appendages as possible, which he might preserve in brine, as also a whole and sound male, as I supposed that I should be back in Dyrefjord about at the time, when he should return to deliver his next cargo of liver, this, as is well known, being the only part of the animal made use of. Circumstances however would that the Ingolf did not return on the Dyrefjord until the beginning of August, and so I did not find the man again. But I found at the manufactory a great deal of pairs of ventrals in brine, all with the appendages and with these in different stages of development, together with a whole male, the last the fisherman had caught; he had during the whole time very carefully kept the last caught male for preservation, and had come on the Dyrefjord with a quite sound specimen, which was also the very smallest he had got; but as I did not return in due time, also this specimen was put into brine. Apparently everything had kept very well by this mode of preservation, the fins at all events excellently; but by the dissection of the whole Shark it soon became apparent that all the internal organs were sadly damaged: the kidneys and the internal reproductive organs were completely disorganized, so that nothing whatever was to be recognised; not even the renal ducts that use to be rather resistant, were to be traced at all. I was thus disappointed in my

---

[1] Collett however states that specimens of a length of about 2 ft. sometimes have been obtained; probably newborn youngs. Meddelelser om Norges Fiske i Aarene 1879–83. p. 118. Nyt Mag. f. Naturv. vol. 29.

hope of being able to give a good account of the structure of these organs, and must be content to give informations of the external copulatory organs.

The whole Shark was about 8 ft. (2^m 50^{cm}) in length, and its ventrals, as also their appendages, were smaller than any of the other cut off ventrals and their appendages, which latter were also much more developed; unfortunately no statement of the length of the respective animals was given. But if we start from the supposition, which I think most likely, that the ventral proper grows in proportion to the animal itself, we can with some certainty calculate the size of the animals, to which the cut off fins have belonged: and judged by that they have all been large animals between 3 and 5 metres, the largest at all events upwards of 6 ells (Danish).

I am not able to decide with perfect certainty, if any of the obtained ventrals have the appendage so large and developed, as it possibly can be; but at all events these organs are so far developed in the largest specimens that they will scarcely change their structure in any considerable degree, even if they become somewhat longer. In the largest fins the free end of the copulatory organ reaches about 5^{cm} farther back than the point of the fin-membrane itself; in the somewhat smaller ones 3–4^{cm}, and in two a little smaller still about 1^{cm} behind the point of the fin. In the smallest specimen finally (the above mentioned animal 2^m 50^{cm} long) the point of the ventral on the contrary reaches 2 3^{cm} farther back than the point of the appendage. Between this last specimen and the immediately preceding the above mentioned specimen of the museum (which however is partly skeletonized) may be placed with regard to size and development. Here accordingly we have a series showing the stages in the growth of these organs, well known from the other Sharks, from small short rudiments, shorter than the ventral itself, to a more or less considerable length beyond the inner edge of the ventral. Thus every idea of the Greenland Shark differing from other Sharks in only possessing rudimentary ventral appendages must be dropped.

About the remaining external features of the organ I shall confine myself to state, that its whole dorsal surface (i. e. the surface which in the natural position is in contact with the ventral side of the body) as well as the adjoining part of the fin itself is quite naked and smooth without dermal teeth, which is also the case with the medial surface, where those of the same part are in contact, while the ventral surface (as in the remainder of the fin) is clothed with dermal teeth, however more sparsely and sparingly towards the point, the outermost part of which is naked and quite soft. Otherwise these organs are in their developed state stiff and hard on account of the strong internal skeleton. On the lateral side of the end is felt through the skin a particularly hard and movable part of the skeleton, and in most of the specimens this part is naked and appears as a pointed, polished thorn or spine. I can however assert with certainty that in all the specimens, I have brought home, it has only been laid bare by the skin on the spot being torn; it is also seen quite covered in the right clasper of one of the largest specimens. I suppose, however, that before the member comes into function, or at the function, this spine is uncovered; in fully developed appendages of *Acanthias* and *Spinax* at all events both the corresponding part and one or two more parts of the skeleton protrude naked, uncovered by the integument; and in the circumstance that in all these fins the spine surely only has been set free by damage or by bad preservation, I find a positive intimation of their appendages not yet having reached their greatest development. This spine is still plainly felt in the somewhat smaller fins, excepting the two smallest; in these evidently it has not yet been calcified, no more than most of the other parts of the skeleton, characterizing the end or terminal part of the developed organ; therefore these small appendages are upon the whole rather soft to the feeling and with flexible ends.

The form of the developed appendage is straight, somewhat dorso-ventrally flattened; a distinction may be made between the considerably longer proximal part, which might be called the shaft, and the short distal part, the terminal part, which is free of the fin, and, as will be more particularly bespoken hereafter, possesses a certain limited mobility; the largest breadth is found immediately before the terminal part; on the dorsal side, somewhat nearer to the lateral than to the medial edge, is seen the peculiar cleft, the appendix-slit, which is found in all Selachians; it reaches to the posterior end of the member, and leads in the free part of this into a deep canal, more anteriorly into a glandular bag, which, like a deep pocket, at the base of the appendage goes round to the ventral side of the fin, and here under the skin reaches — according to age and development — a longer or shorter distance towards the pelvis. The inner walls of this bag are smooth, partly pigmented, and from their epithelium is secreted a peculiar fluid, which when coagulated is tallowy, but whose function is not certainly known. This bag, as to its origin, is simply a folding in of the outer skin [1]; it is surrounded with muscles, able to press the secretion into the canal and through the slit to the exterior. The inner (medial) lip of the slit is immovable and cannot be displaced, while the outer (lateral) one till near the terminal part consists of soft tissue, and is therefore easily opened, so that a finger may be introduced into the bag; but at the end of the shaft, immediately before the terminal part, all distension is prevented by the inner skeleton, which is found here, and straightens the slit, so that it becomes very narrow; to the distal side of this straightening, in the terminal part itself, the canal may again be opened, and it will open spontaneously, if the terminal part is bent a little in the ventro-medial direction, in which case the spine will at once erect.

The following measures referring to the largest appendages, may be added:

Length from the anterior border of the cloaca to the terminal point of the appendage 24 26^{cm}.

of the terminal part of the appendage . . . . 5 5.5^{cm}.

Breadth of the appendage before the terminal part . . . . 3.3 4^{cm}.

Length of the slit . . . . 16^{cm}.

of the part outside the fin . . . . 6^{cm}.

Part outside of the point of the fin-membrane . . . . 5^{cm}.

The skeleton (pl. I. fig. 1 -9). The skeleton of the ventral fin in the male consists of 1) the pelvis, 2) the axial part or the stem, which laterally wears 3) the rays, and as a continuation 4) the skeleton of the appendage.

The structure of the pelvis is as commonly in the Sharks, it consisting of an unpaired, somewhat arcuated cartilage, the surface of which is rather slightly calcified; it has the greatest thickness

[1] I have followed its development in embryos of *Acanthias*, as has also been done by Petri: Die Copulationsorgane der Plagiostomen. Zeitschr. f. wiss. Zoologie, vol. 30, 1878.

in the middle, and here projects from the posterior edge a clumsily rounded process. The stem of the ventral articulates by its principal piece, the basale, (pl. I, fig. 1 *B*), with the lateral end of the pelvis, as do also a pair of the foremost rays. The foremost ray (*k*) is always short and big, shaped like the blade of an axe, whose head articulates with the pelvis, the hindmost corner of the blade with two small terminal joints; it bears the second ray, which is accordingly out of connection as well with the stem as with the pelvis; sometimes it is proximally coalesced with *k*. The third ray has pressed so far forward, that it articulates both with the stem and the pelvis. Most of the other rays are more or less straight, cylindric, distally a little flattened (especially in the foremost ones); the two (less frequently three) hindmost are always somewhat bent, so that the convexity turns dorsally, owing to the fact, that the glandular bag from the dorsal side passes under them to the ventral side of the fin. These two hindmost rays are often more or less united, sometimes almost quite coalesced. The foremost rays (more than half of them) have three joints, then follow some (3) with two joints, and the last (3) are never jointed. The number of rays varies from 12—16[1]); commonly one fin of the same pair has a ray more than the other, and a rather considerable variation is found in the more special relations of the rays, in their mutual coalescing[2], their articulation, and distal dichotomy; sometimes an extra ray is inserted, not reaching the stem; such extra rays have not been counted in the numbers given, and they do not occur symmetrically in both fins. Such variations are also known in other Sharks[3]), and I shall not here enter into further details, as they are of no importance for the examination in question.

The stem consists of 1) a large and big principal piece, *Basale metapterygii* (*B*), to which most of the rays are attached; its inner edge is almost straight, only slightly concave, the outer edge is convex; 2) a short piece (*b₁*) directly continuing the foregoing; 3) generally is on the medial side inserted, as it were intercalated, a little cuneiform piece (*b₂*). The piece *b₁* bears the two hindmost rays, so that the last but one is articulated at its proximal extremity, and here also touches the basale, the last at its distal extremity, where it has also a little articular surface with the proximal end of the stem of the appendage. Finally is found 4) a rather considerable piece (*β*) placed on the dorsal side of the stem in such a way, that it is proximally connected with the latero-dorsal corner of the basale by a little articular surface, and distally by a longer, obliquely placed articular surface with the latero-dorsal edge of the anterior end of the appendix-stem (fig. 2 at *x*). This piece *β* is rather thick, dorso-ventrally somewhat flattened, has a convex medial edge, and a straight lateral edge; posteriorly it is somewhat more pointed than anteriorly; the foremost part of the convex edge is connected with the dorsal side of the piece *b₂*; it has no articulation at all with any of the rays[4]. Between the lateral corner of *b₁*, *β*, and the appendixstem 5) a little piece *b₃* is sometimes intercalated.

Then follows 6) the appendixskeleton. Its chief piece (tab. I, fig. 1 *b*, fig. 2, 3) evidently belongs to the stem, and is placed in immediate continuation of the foregoing pieces, with

---

[1]) In two females I have found the number respectively 15 17 and 16 17 on the two sides.
[2]) In one specimen separate, independent pieces of cartilage have been developed; they are placed across, and near the outer end of the rays they connect two and two of these.
[3]) Comp. Gegenbaur: Ueber das Skelet der Gliedmaassen der Wirbelthiere im Allgemeinen und der Hintergliedmaassen der Selachier insbesondere. Jen. Zeitschr. 5 Bd., 1870, p. 435 seq.
[4]) By the choice of the letter-marks I have intended to point out, that all these parts belong to the stem-skeleton.

which it forms an -- to be sure very obtuse — angle. In a fully developed skeleton the chief piece is longer than the basale; in the largest specimens at hand the ratio is: $\frac{H}{b} = $ c. $\frac{3}{4}$; on the medial side it is rounded, in the foremost third part somewhat dorso-ventrally flattened; the lateral surface (*l*) is more or less distinctly bounded from the other surfaces; it is only in the fore part somewhat rounded, posteriorly it is flattened, and the hindmost part is somewhat hollow; on the dorsal side this lateral surface is in the whole length of the piece sharply limited by a thin, elevated, hard calcified ridge (fig. 2, 3, *Rd*), anteriorly beginning as quite low, posteriorly becoming higher and higher, as well as thicker, and bearing in the posterior half an edge, folded to the dorsal side, irregularly indented, and collarlike; on the ventral side (see fig. 3) the lateral surface is in the greater part of its extent much more indistinctly bounded by an evenly rounded eminence, which is not harder than the common surface; in the posterior part, however, rises rather suddenly a short, calcified, strong ridge or plate, which in the shape of a large foliaceous process folds over to the dorsal side, where it approaches rather near to the opposite edge (fig. 2, 3, *Rr*). The free edge of this folded process is thickened, and irregularly rugged. The described elevated ridges or plates in connection with the flatly hollowed hindmost part of the lateral surface forms the place of part of the appendix-slit or the excretory duct of the gland-bag; these hard parts of the skeleton it is, that, as mentioned on p. 7, prevent a distension of the appendix-slit.

Immediately behind the end of these calcified ridges the chief piece continues as a thin, round, finger-shaped elongation, the end-style (fig. 1, 2, 3, *g*); it is soft, or at all events at its base quite devoid of calcification, while farther out a slight surface-calcification may be found. Else the chief piece is everywhere calcified on the surface (being anteriorly somewhat rough for the attachment of the muscles), and more calcified than the basale and the rays, but the above mentioned ridges (*Rd*, *Rr*) are completely calcified and hard. When such a chief piece is dried, these ridges therefore will not shrink, but rise distinctly as independent parts. By a close examination of an undried chief piece the boundary lines of these calcified side-parts may also be distinguished, and thus we shall arrive at the same result: the chief piece is composed of three parts, viz. the appendix-stem (*b*), posteriorly becoming lanceolate, medio-laterally compressed, and ending as a slender, thin, (at the base) uncalcified end-style, and two calcified marginal cartilages, one long, slender, dorsal, the other shorter, broader, ventral (*Rd*, *Rr*).

To this chief piece are attached a number of terminal pieces, more or less movably joined to each other and to the chief piece. Of these pieces two join the posterior borders of the marginal cartilages and the end-style of the stem, and form, as a kind of continuation of the marginal cartilages, the dorsal (dorso-medial), and ventral (ventro-lateral) borders of the hinder part of the appendix-slit; these two pieces are here called respectively the dorsal and the ventral terminal piece (*Td*, *Tr*).

The dorsal piece (fig. 1 *Td*, fig. 4, 5) is the smaller one; it tapers to both ends, most to the posterior; on the exterior (medially) it is somewhat rounded, with a sharp lateral edge, a little denticulated, towards the appendix-slit slightly hollow in the foremost two third parts; the thick medial edge is by means of connective tissue closely connected with the end-style, the anterior end with the dorsal marginal cartilage. It is completely calcified, and the surface, especially towards the terminal end, is rugged and rough.

The ventral terminal piece (fig. 1 *Tv*, fig. 6, 7) is considerably larger; the surface towards the appendix-slit is deeply hollow like a trough, the external, ventral, surface is rounded, and has laterally a winglike, sharp process; it is also completely calcified, and a great part of the surface is irregularly furrowed and rugged. The one anterior edge of the trough articulates with the ventral marginal cartilage, by the inner, ventral, edge it is connected with the style.

Between this piece and the overlapping plate of the ventral marginal cartilage is seen a third terminal piece (fig. 1, *Tj*), the thorn or spine (fig. 8, 9). It is, like the other pieces, quite hard, and the proximal end is somewhat head-shaped with a smooth surface, almost like an articular surface; else it is for a great part very irregularly rugged and furrowed, but the outermost point is glossy and smooth, dentine-like; the whole thorn is longitudinally somewhat twisted.

Besides these fully developed terminal pieces indications of two more are to be seen, viz. a thin, narrow lamella, only calcified in spots, joins the lateral edge of *Td*, and supports the edge of the dorsal lip of the appendix-slit; anteriorly it reaches somewhat beyond *Td*; this indicated piece is here designated as *Td*, (comp. pl. V, fig. 61, 62); the second piece is a very firm and strong fibrous tissue, joined to the anterior dorsal edge of the piece *Tv*, and without distinct borders merging into the aponeurotic covering, connecting the thorn, the piece *Tv*, and the overlapping plate of the marginal cartilage, and serving for insertion of part of the muscles (see pl. V, fig. 61, 62, *Tv₂*); in this latter piece a calcification has commenced, indicating perhaps, that it might become a separate terminal piece, which I shall designate as *Tv₂* (comp. other Plagiostomes for inst. *Spinax*). As these two last mentioned pieces are, as it were, still developing, I suppose, that even the most developed of the appendices in hand cannot, in a stricter sense, be said to be full grown yet; but as the piece *Tv₂* also in some other Sharks (f. inst. *Acanthias*) is found only indicated and uncalcified, even in quite developed appendices, my supposition is not quite reliable.

The whole of this terminal skeleton, composed of the terminal pieces and the end-style of the stem, is movable to a certain degree; as to further details on this point the reader is referred to p. 14.

By examining the appendix-skeleton in the earliest stages of development we find that originally it is composed of only one single piece, being that, which above is termed the appendix-stem. This (in the specimen from Iceland, 2ᵐ 50ᶜᵐ long, and in the specimen from the Zoological Museum, 9 ft. long) is still quite soft, shorter than the basale, anteriorly rounded, posteriorly lanceolate, the edges of the lancet being placed almost dorsally and ventrally, and ends as a thin style (see fig. 2 in the text p. 19); thus mainly rendering the form of the chief piece minus the marginal cartilages. Of these latter as well as of the terminal pieces no trace is found. In somewhat more advanced stages, where the appendix-stem is as long as, or a little longer than the basale, the three terminal pieces and especially the thorn are very well to be distinguished, while the marginal cartilages still are absent, or, at all events, in the fibrous tissues, occupying their place, no calcification or distinct bordering of such cartilages is to be found (not even of the overlapping plate). In still a little more advanced stages also the marginal cartilages are found in the same shape and with the same bordering as in the most developed, but the boundary lines between them and the stem are much more distinctly marked; they are calcified, but are still soft enough to permit of easy cutting;

to the naked eye the section shows a particular fibrous texture (as in sections of the terminal pieces), and a whitish colour, distinguishing it distinctly from a section of the appendix-stem or any other part of the skeleton proper, for inst. a ray or the basale, the surface of which will be hyaline. From these developmental facts it will appear with all desirable distinctness, that the marginal carti-lages and the terminal pieces are secondary parts of the skeleton, developed in the tissues surrounding the primary skeleton, properly so called. Thus of the appendix-skeleton only the appendix-stem, the piece *b*, belongs to the primary skeleton.

To resume what is said about the appendix-skeleton in the Greenland Shark:

The appendix-skeleton consists of a chief piece and terminal pieces movably connected with it; the chief piece is formed by the coalescing of the appendix-stem with two secondary calcified cartilages, the marginal cartilages; the appendix-stem belongs to the primordial axial skeleton of the ventral fin, being the terminal joint the extremity of which remains soft; the terminal pieces are all secondary calcified cartilages.

The muscular system (pl. V, fig. 58 to 62) follows the type, which has been described in *Acanthias* by v. Davidoff[*]; this type, however, has been founded on the structure of the ventrals of the female; the rather considerable differences from it are due to the copulatory appendages, for the special use of which special muscles have to be developed. Distinction may be made between: I) The fin muscles proper, and II) the muscles of the appendage; as, however, some of the former spread over part of the appendage, this distinction cannot be made quite distinct.

I. In the fin-muscles proper may be distinguished, as v. Davidoff and the earlier authors do, between the muscles of the ventral and those of the dorsal side; they are anta-gonistic, the former adducting the fin, and removing it from the abdomen, the latter abducting the fin, and pressing it against the abdomen.

1) The ventral muscles of the fin consist of a) a medial muscular mass, chiefly reaching from the pelvis to the stem-skeleton of the ventral, with laterally and obliquely-posteriorly directed bundles of fibres, and b) a lateral mass, the muscles of the rays, issuing from the stem-skeleton, and following the rays to the fin-membrane.

a) This powerful group of muscles (pl. V, fig. 58—61, *A* and *E*) in so far does not wholly be-long to the ventral side, as, besides forming the medial edge of the fin, it is also seen on the dorsal side. Looking first at its ventral side we find its origin covering almost the whole ventral surface of the pelvis; between the fin-muscles of the two sides only a triangular piece of the pelvis is to be seen in the middle anteriorly, from the top of which a narrow uncovered streak runs backward to the end of the above (p. 8) described process; from this issues further backward in the linea alba an aponeurotic streak (fig. 58, *s*), which continues the pelvis, and serves as attachment for part of the same muscular mass. The superficial ventral part is for the greater part composed of distinct bundles of muscles, enveloped in rather firm sheaths of connective tissue, and mostly corresponding in number and direction with the muscles of the rays; but this composition of isolated bundles is effaced anteriorly-laterally and posteriorly-medially.

Anteriorly the fibres running obliquely from the pelvis towards the outer margin of the fin

*) Beiträge zur vergleichenden Anatomie der hinteren Gliedmasse der Fische. Morphol. Jahrbuch. 5 Bd., 1879, p. 454 seq.

form a rather solid mass spreading from the fore edge of the pelvis over the broad ray R. The foremost of the following distinct bundles of muscles cross the ventral surface of the basale reaching as far
as to the horny filaments of the fin-membrane, ending here in a tendinous mass; the following bundles
only reach to the basale where they are inserted with tendinous ends, from which tendinous part the ray-
muscles originate as a prolongation — however, when we look farther backwards, with a distinct inter-
position of a narrow stripe of the basale. Between the said foremost bundles, continuing immediately in
the ray-muscles, and those attached to the basale, a gradual transition is found, a tendinous part in
the superficial layer of the bundles being inserted on the place of transition.

The hindmost and medial part of the muscle A is not composed of isolated bundles, but its
fibres running rather straightly backwards form a solid mass, inserted on the distal end of the basale,
on the pieces b_1, and b_2, and on the proximal end of the chief piece of the appendage (b).

The whole muscular mass, as mentioned, is of a considerable thickness; its deeper part which
is also seen from the dorsal side, is not divided into separate bundles; this deeper, more dorsal, part
originates from the rounded posterior surface of the pelvis, and even reaches to its dorsal surface; it
is inserted along the medial side of the basale and the following joints inside the insertion of the
described superficial ventral layer.

With this muscle A is closely connected another (pl. V, fig. 59 and 61, E), chiefly seen from
the dorsal side. It originates on the medial side of the basale, a little before the middle. its fibres
crossing those of the muscle A, and spreading over the appendage; as above the knee of the latter
the fibres run obliquely across the medial edge of the fin and on to the ventral side, part of the edge of
this muscle will consequently be discernible on this side (pl. V, fig. 58 and 60 E). It is spread like a
cloak over the chief muscle (D) of the appendage forming a rather thin plate and growing thinner from
the ventro-medial edge laterally (cp. fig. 1 in the text); its fibres are attached, partly along the narrow
ridge, formed by the dorsal marginal cartilage along the appendix-slit (fig. 61 d) partly, distally, to a
thin, firm aponeurosis (fig. 61 m), covering the muscle D, and attached to the elevated distal part of
the dorsal marginal cartilage (Rd). In somewhat older animals with well developed appendages this
muscle E is as well proximally as distally distinctly separate; in young animals, however, with
only little developed appendices (fig. 59) the distal part is still very distinctly marked, but the proximal
part is less sharply separated from the large muscular mass A; numerous bundles coming from the
pelvis and the aponeurotic streak s unite with those from the basale, and numerous bundles from the
basale run over among the former and reach to the proximal end of the appendix-stem.

The above described muscular group consisting of the muscles A and E, will, according to
circumstances, be able to act in two different ways; these muscles will, when the antagonists of the
dorsal side are not contracted, move the fin from the abdomen, and at the same time draw its inner
edge towards the median line, thus moving the two fins towards each other; and when the dorsal
antagonists act on the fin, they will move the appendix only, towards the median line, thus
acting as extensors for the appendix; the latter action will be facilitated by the muscle E acting
rather distally on the appendix (an effect as to the opening of the appendix-slit is of course out of

the question). Consequently I design the large chief muscle *A* as *Musculus adductor (el depressor) pinnæ (et appendicis)*, the muscle *E* as *Musc. extensor (appendicis)*[1].

b. The ventral muscular system of the rays (fig. 58, 60, *Ra*) is composed of distinctly separated bundles of fibres, or independent muscles in number corresponding with the rays they follow; only anteriorly the independence of the ray-muscles, as mentioned above, is concealed by coalescence with the lateral bundles of *Musc. adductor*, coming from the pelvis. The ray-muscles originate on the ventral surface of the basale and the piece b_2, and run laterally backwards in an oblique direction, each following its ray, but without reaching the end of it; they only reach the horny filaments (the two layers of which comprise a rather considerable part of the lateral ends of the rays) and here pass into tendinous tissue. The hindmost ray-muscle is rudimentary; it does not originate on the stem-skeleton, but on the last ray but two, and passes to the last but one and on to the fin-membrane.

2) The dorsal muscular system of the fin proper (pl. V, fig. 59) is composed of a) a superficial part originating from the lateral muscles of the body, and b) a deeper-lying part originating from the stem-skeleton.

a) On a part of the body, corresponding in length to the connection between the body and the fin, a system of distinct muscular bundles (*O*) originate in the aponeurosis covering the lateral muscles of the body, and run obliquely outward and backward to the horny filaments, where they pass into tendinous tissue; thus their outward border corresponds to that of the ray-muscles on the ventral side, being considerably distant from the ends of the rays. The hindmost of these bundles are directed straight backwards, corresponding to the direction of the last of the rays. Furthermore from the inner side, the side towards the muscles of the body, of the said system some bundles of fibres (*O'*) originate running obliquely backward and inward, and attached to the hindmost half of the basale and to the dorsal piece *β*; thus the whole system originating from the lateral muscles, is, as to the hinder half, arranged in a feather-like or fanshaped way.

b. Quite covered by the superficial layer just described the deeper layer of the dorsal ray-muscles (fig. 59, *Ra*) is found. These muscles originate from the dorso-lateral side of the basale and of the piece b_2 as well as from *β*, and are seen as distinct bundles corresponding in their number and direction to the rays; they pass into tendinous tissue immediately before the lateral ends of the bundles of the superficial layer, so that the latter reach a little way farther on the rays. However, these two layers are not quite sharply separated, bundles of fibres from the superficial layer reaching to the deeper, and connecting with it; on the hindmost fin-rays the bundles of the deeper layer cross those of the superficial one, this latter spreading in a fanshaped way from the attachment to the body.

II. Besides the described separate parts of the fin muscles connected with the appendix

---

[1] The muscle which in *Acanthias* and other Selachians corresponds to the muscle *A*, is by Petri called: *Flexor pterygopodii*; but there are several objections to this name. Firstly, the muscle does not only act on the appendix, but on the whole fin (it is also found in the females, and next it cannot well be called the flexor of the appendix, as it is more properly to be regarded as the extensor. The flexion of the appendix is, I suppose, effected by means of the *M. compressor sacci*, the muscle of the glandular bag (fig. 58, 61, *S*), of which more hereafter, together with the muscular layer originating on the body itself (fig. 59 and 61, *O*).

other muscles are found, more especially belonging to this organ, it being inside the skin quite surrounded by muscles except the terminal part. In this muscular system may naturally be distinguished between: 1) The muscles of the chief piece, and 2) those of the glandular bag.

1) The first part (pl. V, fig. 58—62, *D*) is composed of one single muscle wrapping in a cloak-like manner the whole of the chief piece from the dorsal marginal cartilage to the ventral one, and to the rounded edge formed by the appendix-stem itself along its lateral surface above this short marginal cartilage; the part of the chief piece situated between these bounds, the lateral surface is for the greater part covered by the muscles of the glandular bag (see fig. 1). The large muscle *D* is thickest along the medial side of the appendix, and is chiefly composed of longitudinal fibres arising from the whole length of the chief piece; from the foremost part of this, below the knee, arise some specially powerful bundles, and consequently this part of the surface of the skeleton is very rugged; also from the lateral edges arise numerous fibres and bundles, and distally several bundles come from the covering aponeurosis *a* (see pl. V, fig. 61). Corresponding to the form of the appendix-stem this muscle tapers distally, and its hindmost fibres reach to the base of the style. It is inserted in the firm aponeurosis covering the marginal cartilages and the whole terminal part, and thus it acts on the style and the two terminal pieces *Td* and *Tr*. In contracting it bends the style medially forward at an obtuse angle to the chief piece, whereby the two terminal pieces are also moved; at the same time the thorn is erected on account of its connection with the other terminal pieces, especially *Tr*, and stands out laterally; as a consequence the distal part of the appendix-slit situated between these movable pieces, is dilated to a rather considerable degree. I therefore (like Petri) design this muscle as *M. dilatator*.

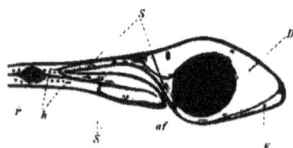

Fig. 1.

Part of a transverse section through the appendage of the Greenland Shark (about 26mm behind the beginning of the appendix-slit). *b* the appendix-stem; *D M. dilatator*; *E M. extensor*; *S M. compressor*; *af* the appendix-slit; *r* a ray; *h* horny filaments.

2) Among the muscles of the glandular bag I do not only class a) the muscles immediately wrapping this organ, but also b) some portions (fig. 61, 62, *S*) arising from the hindmost rays, and forming, in my opinion, with the glandular bag an insolvable whole, only artificially to be detached from it. The glandular bag, as I understand it, has its origin from an invagination of the skin into a muscular mass laterally covering the stem-skeleton in the appendix; by the further growth of this invagination on to the ventral side of the fin part of the muscular mass was brought along as a kind of wrapping of the bag and developing further together with it. Consequently this wrapping cannot be regarded as dermal muscles but belongs to the skeletal muscles; it is also composed of quite the same striated fibres as these; its original relation to the stem-skeleton may, in the fully developed organ, be seen in the still existing attachment along the lateral surface of the appendix-stem (see the transverse section, fig. 1 in the text).

a. The glandular bag (pl. V, fig. 58, 60 *S*) is seen on the ventral side of the fin, where it reaches forward covering a smaller or larger part of the ray-muscles, according to the development of the whole appendage; while in the youngest specimens it only reaches very little beyond the knee between the stem and the chief piece of the appendix (cf. fig. 58), in the most developed it reaches almost half way towards the pelvis (cp. fig. 60). As the glandular bag in most of the other Sharks, which I have examined, reaches still further, generally even far beyond the pelvis, there is reason to suppose that in none of the ventrals of the Greenland Shark in hand the whole copulatory organ has reached the greatest development, which was already intimated by the description of the skeleton of the terminal part.

The connective tissue, investing the muscle-sheath of the glandular bag, is continued on all the specimens as a very thin membrane between the skin and the ray-muscles almost to the pelvis; this membrane may easily be separated as well from the skin as from the muscles, but in the specimens in hand it (perhaps as a consequence of the preservation in brine) is very fragile; it contains no striated muscular fibres.

While the dorsal muscular wall of the glandular bag has no intimate connection at all with the part of the fin before the knee – only a loose, soft connective tissue here joining the bag to the ray-muscles – it is otherwise at the proximal end of the chief piece, part of the muscles of the bag being inserted on the lateral surface of this part of the skeleton, covering it wholly, and following it quite down to the terminal part; other fibres attach to the last ray along its medial edge; and some fibres arising from this spot and from the ventral surface of the two last rays, pass into the dorsal muscular wall of the glandular bag and continue it to the ventral marginal cartilage, where they attach to the connective tissue of its inner side.

The direction of the fibres of the dorsal muscular wall of the bag otherwise corresponds to that in the ventral wall; as shown in fig. 60, the fibres radiate from the point, where the connection with the skeleton anteriorly ceases; along the medial side they run almost in a parallel direction with the axis of the bag and the appendage, but else on the broader part of the bag they spread in a fanshaped manner to the lateral edge; on the hindmost narrow part they run entirely straight backward, and here a few bundles pass into *M. dilatator*. This arrangement agrees very well with that, which fibres originally directed from before backwards, might be supposed to get by being pressed out of their position by an invagination protruding from the region between • • in fig. 60. A separation of the muscular wall of the bag into two distinct layers is quite out of the question. With regard to *Acanthias* Petri (l. c. p. 316) has stated that the muscular wall of the bag consists of two layers, an outer one of circular muscles, and an inner one of longitudinal muscles; a separation and arrangement of such a kind however, is not found in *Acanthias*, any more than in *Spinax* or the Greenland Shark. Neither can I admit that the words of Petri (l. c. p. 317) are correct: Die Muskelschicht der Drüse wird nicht mit eingestülpt, sondern sie differenzirt sich allmälich aus der Bindegewebsschicht nach der Einstülpung. (Cp. also l. c. p. 328). In my opinion, as before has been shown, it admits of no doubt that the muscles of the bag are simply borrowed from the original muscular system of the skeleton[1]; in the earliest stages of *Acanthias* — male embryos of a length of 15^{cm} — which I have been able to examine in this respect, the muscles around the rudiment of the glandular bag are already as distinct as those surrounding the stem of the chief piece, and the muscular layer of

[1] This is corroborated with particular plainness by the arrangement in the Holocephales.

the bag has already been pressed towards the ventral side as well as the other surrounding layers of tissue.

b. Intimately connected with the other muscles of the glandular bag is found a powerful muscle (pl. V, fig. 59, 61 and 62, *S*), seen on the dorsal side, where it forms the lateral lip of the long slit (*af*), which is the entrance to the bag. It takes its origin from the two hindmost rays (sometimes also having bundles from the last but two) as also from the lateral surface of the piece *β*, covered by the superficial layer (*O*) coming from the muscles of the body; it is inserted in the tendinous tissue passing over the head of the thorn (*T₃*), and firmly connected with the proximal end of the terminal piece *Tv*, especially with its edge; in this tissue is found imbedded several firm, fibrous portions, which partly calcify, and probably   in more developed stages — form a separate piece (*Tv₁*). In the hindmost part this muscle is completely fused with the distal part of the muscles of the glandular bag, and anteriorly it forms a whole with the above mentioned bundles of the dorsal wall of the bag, which arise from the ventral side of the two hindmost rays; in the interspace a kind of separation is effected by the attaching of the fin-membrane, the connective tissue of which wedges in between the lip muscle and the wall of the bag itself. This muscle acts antagonistically to *M. dilatator*, which in a preparation is easily seen by pulling it; thus when *M. dilatator* by contracting has dilated the groove between the terminal pieces, as described above, and the thorn stands out, the contraction of this outer lip-muscle of the appendix-slit will again straighten the groove by especially acting on the piece *Tv*, and at the same time carry back the thorn , so that it will lie against the piece *Tv*.

I find the same muscle in all other Plagiostomes, but in very different stages of development (cp. the following). Petri has mentioned it in *Acanthias*, but as *M. levator* of the thorn (l. c. fig. 5, *B, C, F, ml*); he says: Er inserirt sich hinten vermittels eines starken, sehnigen Bandes am vorderen Theil des Spornes (the thorn) und hat allein die Aufgabe diesen zu heben. This, however, is quite incorrect: it is not inserted on the thorn, even if its tendon of course by looser tissue is connected with the proximal part of the latter, but on the piece *Tv* (*b'''* in the figures of Petri), of which piece Petri's interpretation is quite wrong (cp. the following); and it does not assist the following; *M. dilatator*, nor raises the thorn, but it counteracts the *M. dilatator*, and thereby becomes a *M. depressor* of the thorn!

The carrying back to the position of rest of the terminal pieces is in the Greenland Shark and *Acanthias* not exclusively brought about by an elastic reaction of the tissues between the firm parts of the skeleton, as asserted by Petri (l. c. p. 303), but this reaction, which certainly exists, is also supported by the action of muscles belonging to the glandular bag, or, at all events, forming part of its muscular system. Taking it for granted that the appendix genitalis by the copulation is really introduced into the cloaca of the female, I imagine the following act to take place: the appendix is guided and brought into the cloaca by means of the muscles belonging to and arising from the fin-muscles proper; next the *M. dilatator* will come into function, and, by its dilating the terminal parts, fix the appendix in the cloaca, and then the muscles of the glandular bag will evacuate its contents into the furrowshaped, in the appendix itself situated part, the walls of which at the same moment will contract, at the same time ejecting the secretion and letting go the firm hold of the apppendix. As I think the chief action of the muscular wall of the glandular bag to be the ejection of the secretion, I design it — including the described outer lip-muscle of the appendix-slit   as *Musculus compressor (sacci)*.

II.
The Ventral Appendages in other Selachians.

For comparison with the facts found in the Greenland Shark, I have examined as many other forms of Selachians, as I have been able to get the material for, being soon convinced that the representations, hitherto found in the literature, gave only a rather incomplete insight into the structure of these organs, and only to a small degree were to be used comparatively.

The greater part of my material has consisted of well preserved ventrals, a less part only of skeleton parts, dried or preserved in spirit, which the director of the collection of Vertebrata of the Zoological Museum, Professor Lütken, has been kind enough to place at my disposal. The following description has been divided into three parts of very different extent, of which the first will give a short general account of the copulatory appendages in the Selachians in general, the second a more particular description of the forms, on which this general account has been based, and the third will as a conclusion contain some short remarks as to what for the present may be regarded as tolerably certain concerning the function of these organs. That the particular description will treat more of the skeleton and less of the muscles is occasioned by the relatively small variation of the latter.

1. A General View of the Copulatory Appendages in the Selachians.

As to the **outer form**, the same outline is found in the copulatory appendages of all Selachians: it is always the inner part of the fin which is prolonged, and formed into an appendage, and this appendage may be more or less free of the fin-membrane; it is most separated in the Holocephales, least so in some Sharks; it always consists of a, longer or shorter, proximal part, the shaft, and a, generally shorter, distal part, the terminal part, this latter being always free of the fin-membrane, and (at all events in the Plagiostomes) possessed of a certain mobility.

On the dorsal side of the appendage, sometimes, however, quite laterally, a deep furrow or slit, the appendix-slit runs longitudinally, to the posterior end; the edges or lips of this slit can always be opened, at least in two places, viz. at the foremost beginning of the slit in the shaft, and behind in the terminal part; frequently the slit can be widened in a considerable part of the shaft (*Somniosus, Acanthias, Spinax, a. o.*); there is, however, always a part of the slit, in which widening is prevented by the inner skeleton, or where the lips cannot at all be separated, or sometimes even may be coalesced (the latter in *Scyllium* and *Pristiurus*); the part of the slit situated in the terminal part can (at all events in all Plagiostomes) be widened by muscular action, and again narrowed by elastic reaction, sometimes assisted by muscular action. The appendix-slit is the duct of a glandular bag which is surrounded by muscles, and in all Plagiostomes with its greater part situated on the ventral side of the fin, under the skin, but in the Holocephales, where it has only been little developed, limited to the appendix-shaft.

The skeleton of the appendage belongs always to the axial stem of the fin-skeleton [1]; among the rays (in the Plagiostomes, not in the Holocephales) only the hindmost, most frequently the two hindmost, are of importance as serving as attachment for part of the appendix-muscles (those of the glandular bag); as a consequence these rays have been somewhat bent, with the convexity turned dorsally; the two hindmost are often partly, sometimes quite coalesced.

With the primary skeletal parts, developed from the fin-stem, join, in the Plagiostomes, several very differently shaped, calcified, secondary skeletal pieces, developed in the connective tissue, surrounding the original, primary skeleton. These secondary pieces show, especially in the terminal part, a considerable variation, both as to form and number, and the different genera, or even species, may present rather important differences; but everywhere may be established the same fundamental type that has been pointed out in the Greenland Shark.

In the Plagiostomes the primary skeleton consists of: a large basale (B), and in continuation of this one or more (until a number of four, *Rhinobatus*) shorter pieces (b_1, b_2, etc.), and finally a terminal joint, the appendix-stem (b); this latter is always long, often considerably longer than the other parts of the stem taken together. To these pieces must be reckoned one more, β, placed dorsally, parallel to the short stem-pieces b_1, b_2 etc.; most frequently it connects the basale with the appendix-stem, but sometimes it does not reach the basale anteriorly, and is then connected with b_1; in *Rhina* it is rudimentary, and only connects the last joint with the appendix-stem; in *Narcine* it seems to be wanting.

In quite young males of Plagiostomes (cp. fig. 2 in the text), even in embryos, all these primary skeletal parts are already found; during the growth the terminal joint, the appendix-stem, is prolonged, growing much more than the other parts, and calcifying to some degree in the surface (often to a higher degree than any other part of the primary skeleton of the fin) always, however, with the exception of the distal terminal part, this often wholly, and at all events at its base remaining soft, and consequently flexible; this part of the appendix-stem I (after its form in the Greenland Shark and many other Sharks) name the end-style (g).

Contemporary with the growth and the calcification the secondary skeletal parts develop around the appendix-stem, first as firm, fibrous parts, calcifying by degrees, and finally very hard; some of them belonging to the terminal part are even shining, polished, and dentine-like; they then rise, more or less naked, through the skin; this applies to one piece in *Somniosus*, *Lamna*, *Selachus*, *Rhinobatus*, *Raja radiata*; to two pieces in *Acanthias*, three (four) in *Spinax* etc.

Two of the secondary skeletal parts are always closely connected with the appendix-stem, and may even quite coalesce with it; these two cartilages form shorter or longer ridges, and are situated, one dorsally, the other ventrally, connected with the appendix-stem in such a way as to form with it the part of the appendix-slit that cannot be widened; they are the two marginal cartilages, the

---

[1] When A. Fritsch (Zool. Anzeiger, vol. 13. 1890, p. 318, and Fauna der Gaskohle etc. Böhmens, vol. 3, 1895) restores the ventral appendages of the fossil Xenacanths as lateral structures, developed from rays, I am convinced that he is wrong, and has misinterpreted the fossils.

dorsal one (*Rd*), and the ventral one (*Rv*); posteriorly they always reach to the end-style, anteriorly more or less forward, commonly not to the same length, and at most to the proximal end of the appendix-stem. Together with this they form the chief piece of the appendix-skeleton, a name used by several earlier authors, who most frequently have not seen that this piece consists of three parts.

The other secondary cartilages, the t e r m i n a l p i e c e s, together with the end-style form the skeleton of the terminal part, and are more or less movably connected mutually, with the marginal cartilages, and with the end-style. The number of terminal pieces may be different, but in all Plagiostomes two are found, one dorsal (*Td*), and one ventral (*Tv*), placed as a kind of movable continuation of the two marginal cartilages, and with their inner edges joining the end-style of the axial piece, which by being bent (ventro-medially) is moved in connection with them; thereby they dorsally withdraw more from each other, and the slit between them is widened. Only in a few cases (*Trygon violacea, Chlamydoselachus*) these two pieces are found alone; in most Sharks a piece *Td₂* is joined to the lateral margin of *Td*, and imbedded together with

this in the dorsal lip of the appendix-slit; often a piece *Tv₂* is in a like manner joined to *Tv*; further is generally found a piece *T₃*, placed ventrally and laterally, and often rising through the skin as a spur or thorn; still more pieces may be developed (especially in *Raja*), but their homologies in the different forms are generally easily pointed out, and are in the special part indicated by the letters used. Finally may to the terminal pieces proper be joined one or more spurious pieces or c o v e r i n g p i e c e s, enclosing like a shield the terminal pieces, properly so called, on the dorsal side (*d*) or the ventral side (*v*); they are developed in the aponeurosis of the *M. dilatator* bespoken later on, which otherwise wraps the terminal part, and serve as insertion for part of this muscle. Such covering pieces are found in all Rays and in some Sharks (for inst. *Rhina*). As to the abundantly varied structure of the terminal part the reader is referred to the special part; here I shall only add that the simpler forms are generally found in the Sharks, to which may be joined among the Rays *Torpedo, Narcine, Rhinobatus* and *Trygon*, while the most complicated structures are found in the species of *Raja*.

Fig. 2.

Fig. 3.

Fig. 2. *Somniosus microcephalus*, young ♂ (12ᵐ 50ᶜᵐ). The hindmost part of skeleton of left ventral (considerably diminished). The letters as before. ● an intercalated extra-ray.

Fig. 3. *Somn. microcephalus*, ♀. The corresponding part of skeleton of left ventral. *b₁* + *b₂* two coalesced stem-joints; the stippled line indicates the distinction, found between these joints in the right ventral of the same specimen. *b* the terminal joint. Reduction as in fig. 2.

Perhaps it may not be devoid of interest to compare the ventral skeleton of the m a l e with that of the f e m a l e. In this latter we find the stem composed of a large basale and a different number of shorter joints, among which the terminal one has no ray (typically), but often looks like a ray

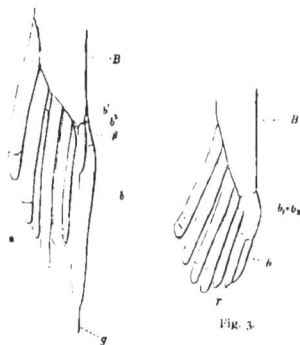

3*

itself being more or less rodshaped. This latter joint, I suppose, is the one that in the male is prolonged and developed into the appendix-stem, which never bears rays; otherwise, however, the number of 'intermediate joints' between the basale and the terminal joint (the appendix-stem in the male) does not always correspond in the two sexes of the same species, and the part of the stem situated distally of the basale seems upon the whole to be rather varying in females of the same species[1]. In the female, as was to be expected, all the secondary skeletal pieces are wanting, but besides those also the piece β of the primary pieces. It is rather difficult to decide with any degree of certainty, how this piece is to be interpreted; perhaps it might be done by following its development. The smallest embryos (of *Acanthias*) that I have had occasion to examine, however, have had this piece quite independent, in the same position, and with the same relations as in the grown animal. This piece, however, has to be considered as belonging, either to the stem, or to the rays, and in the latter case it is, I think, to be regarded as one ray, there never being any mark of a composition of more parts. In several species, as *Trygon*, *Rhinobatus*, it might, as to its form, remind of a ray, which then was to be considered as displaced to a higher level than the others, and turned parallel to the axial stem; in *Trygon* it must be the last, hindmost ray, while in *Rhinobatus* it could not be the last ray, as more real rays follow farther backward; and so on in the other species: if it was to be considered as a ray, it must, in the different species, be a different ray, displaced and transformed. I think it more probable that the piece β belongs to the stem, and has been separated from this by a longitudinal division, which might possibly be occasioned by the development of special muscles for the appendix.

In the Holocephales (see pl. I) all secondary cartilages are wanting in the fin-skeleton; it is only composed of a large basale bearing all the rays, of a short piece b_1, the appendix-stem b, and the dorsal piece β. The walls of the appendix-slit are produced by a kind of rolling-up of the stem-portions b_1, and b, and thus the terminal part is only formed of the hindmost part of the appendix-stem; this latter is rather differently formed in the two genera *Chimæra* and *Callorhynchus* (see the special part).

The appendix-skeleton of the Holocephales accordingly is of a less compound construction than that of the Plagiostomes, and that, as will be seen hereafter, is also the case with the muscular system. This simpler structure evidently in some degree repeats primitive features, but these, on the other hand, are connected with facts, that by no means are primitive, as for inst. the strongly marked separation of the whole organ from the fin proper, the highly specialized form of the primary skeletal parts — against the simpler form in the Plagiostomes (as the simple, rod-like shape of the terminal joint b etc.) , the connection with other, particular copulatory organs, etc.; these things, as well as many other facts

[1] In two specimens of ventrals of female Greenland Sharks I find the structure different in the two sides of the same pair of fins. In the left ventral of one specimen the basale is followed by a long and powerful joint, ($b_1 + b_2$, fig. 3) bearing two rays, and a ray-like little terminal joint b; in the right fin of the same specimen follow after the basale two short joints (the distinction between these is indicated by stippling in fig. 3) b_1, b_2, each bearing one ray, and b_2 also the little ray-like terminal joint b; thus on the left side a coalescing of b_1 and b_2 seems to have taken place. On the left side of the other specimen follows after the basale only one sword-like, compressed piece, taking the place as the terminal joint, and showing in its distal end, which is somewhat flattened, an indication of a longitudinal division; in the right side, on the contrary, the basale is followed by a short joint (b_1) bearing a ray and a compressed terminal joint (b). Consequently, if we suppose a coalescing of b_1 and b on the right side, together with the last ray, we shall arrive at the structure on the left side. As far as I have seen, the female fin-skeleton of *Acanthias* shows similar variations.

in the structure of these animals indicate that the Holacephales by no means occupy a primitive position among the Selachians.

As to the skeleton of the ventral in the female, the basale (in *Chimæra*) has distally only one small, tap-like joint, standing both for the piece b_1, and the appendix-stem (*b*) in the male[1].

What has been given in the earlier literature as to the skeleton of the ventral appendages in Selachians, is generally only isolated descriptions without any real understanding; only Gegenbaur[2] and Petri have compared several forms, but neither of them has been able to recognise a common type. Gegenbaur (l. c. p. 452) has interpreted the terminal pieces as modified rays, but on account of the circumstances in the *Chimæra*, he indicates (p. 456) the possibility that they may be parts separated from the stem-skeleton; he does not know the marginal cartilages, and he has considered several early stages of the skeleton as definitive forms of it. Petri quite correctly has seen that the terminal pieces and marginal cartilages — which latter, however, he has not recognised in all the species he has examined — are secondary structures, and have nothing to do with the rays; the terminal stem-joint itself which I have called the appendix-stem (*b*), he has interpreted correctly in *Raja*, but wrongly in *Acanthias* and *Scyllium*[3] — the only Sharks examined by him — as well as in Torpedo[4] (he has not examined *Chimæra*). Some earlier authors have seen the piece β in some specimens, while it by others has been overlooked, or at all events has not been mentioned. Only Gegenbaur and Petri have sought its origin in a transformation of other skeletal parts of the fin[5]. Gegenbaur does not mention it at all in *Raja*, *Carcharias* and *Scyllium*[6] but in *Heterodontus* and *Acanthias* (l. c. fig. 16 and fig. 19, *b*), and in *Chimæra* (fig. 23 *r'*); in the last named it is interpreted as a ray, but in the two former as belonging to the stem-skeleton[7]; accordingly Gegenbaur has not seen that in *Chimæra* it is the same skeletal piece as in the Plagiostomes. Petri thinks it to be a coalescence of basal parts of rays, being of opinion that it bears rays in *Acanthias* and *Torpedo*; accordingly in his figures he marks it *r'*. This supposition, however, is wrong[8]; I never found rays

---

[1] v. Davidoff (l. c. p. 473, pl. XXIX, fig. 18 *b*) thinks it only to be corresponding to b_1. Unfortunately I have only had occasion to examine skeletonized ventrals of *Chimæra* ?, in which this joint was wanting, so that the fin-stem consisted only of the basale.

[2] Ueber die Modificationen des Skelets der Hintergliedmaassen bei den Männchen der Selachier und Chimären. Jen. Zeitschr. vol. 5. 1870, p. 452.

[3] In these Sharks Petri supposes the stem to end with a long and a short joint; in *Acanthias* as the short terminal joint he has interpreted one of the terminal pieces (my piece *Tr*), in *Scyllium* the soft end-style.

[4] As to *Torpedo* see p. 49.

[5] Bloch, M. E.: Von den vermeinten doppelten Zeugungsgliedern der Rochen und Haye. Schr. der Berl. Gesellschaft Naturf. Freunde, vol. 6, 1785 (*Raja clavata* [= *radiata*]); and: Von den verm. männlichen Gliedern des Dornhayes, ibid. vol. 8, 1788, does not mention this piece in *Acanthias*, but in *Raja*, where he calls it: der vierte Knochen des Schenkels, pl. IX, fig. 1, *o*. Cuvier (Duvernoy); Leçons d'anatomie comparée, 2 Ed., 1846, vol 8, p. 306, designates it as Calcanéum in *Raja*; the same appellation is used, likewise for *Raja*, by Moreau: Hist. nat. des Poissons de la France, vol. 1, 1881, p. 249. I have not found it mentioned by other authors.

[6] Of these three forms G. has only had quite young specimens, in which the secondary pieces had not yet developed. The fault made here by G. viz. to consider this stage as the full-grown state, and accordingly as an especially simple form in these Plagiostomes, has already been corrected by Petri (l. c. p. 293). It is to be supposed, however, that the piece β had been developed in all three forms, as in embryos of *Acanthias* of a length of only 13^{cm} it is already quite distinct and relatively as large as in the full-grown animal.

[7] I am quite unable to understand the place in question (l. c. p. 451 in Gegenbaur; there is a regrettable discrepancy between the letters in the text, and those in the figures, and also. I think, a change of pieces, which makes the whole confused; so much, however, is certain that the piece which in the figures 16 and 19 is marked *b* (my piece β) does not in *Acanthias* bear any ray; it never bears rays at all.

[8] For further details see under *Acanthias* and *Torpedo*. When Petri, to support his construction of this piece as

attached to this piece, but always found it placed at another level than that of the nearest rays, and I take it to be a specially separated part of the stem-skeleton.

The muscular system (see pl. V and VI) does not show the rich variation found in the skeleton, being upon the whole rather uniform, which is a natural consequence of the fact that the part of the skeleton, particularly multifarious both as to the number and form of the single pieces, viz. the terminal part, has no muscles of its own; the muscles (generally) only acting on the terminal part as a whole.

Only the medial side of the fin-muscles has been specially developed in the male; the muscles spreading over the lateral parts of the fin, i. e. the ray-muscles of the upper and lower side, and the dorsal layer originating from the lateral muscles of the body, are chiefly the same in both sexes, and show in the different forms examined so very few differences that I, also in the special part, pass over them.

In the medial muscular system may be distinguished between a more proximal and a distal part, not however strongly separated, especially not so in many Plagiostomes, while in the Holocephales the separation is more distinct, the appendage of the latter being more independent of the fin.

In the Plagiostomes I generally find the same type, as has been described in the Greenland Shark. The proximal part consists of a *Musc. adductor (et depressor) pinnae (et appendicis)* (.l), and a *M. extensor appendicis (E)*. *Musc. adductor* does not in any of the forms examined by me show any separation into an independent, superficial ventral layer, and a deeper, more dorsal one, but forms a whole[1]; the ventral side, however, appears to a great extent separated into single bundles corresponding to the ray-muscles, while the dorsal side shows nothing of the kind. The fibres arise from the pelvis, as well from the ventral, as, though often to a smaller extent, from the dorsal surface, as also from a tendinous stripe prolonging, as it were, the hindmost edge of the pelvis into the median line; they run obliquely-laterally, and are inserted on the basale, on the following joints (b_1, b_2 etc.), and on the proximal end of the appendix-stem; often, however, the superficial medial fibres run on and mingle with the *M. dilatator*. The fibres forming the medial marginal part, run almost straight from before backward, and form always a solid mass not divided into separate bundles; the foremost, lateral parts (as in the Greenland Shark) are coalesced with the deeper-lying ray-muscles.

M. extensor (appendicis) (E) is mostly a rather flat muscle, situated on the dorsal side of the previous one; it originates on the medial side of the basale, often moreover on the pieces b_1, b_2 etc., and is inserted on the appendix-stem, usually at the proximal end, but sometimes farther backward, and the hindmost part of this muscle then spreads in a cloak-like manner over part of *M. dilatator* (comp. the Greenland Shark). This muscle generally is very distinct, already in quite young animals with undeveloped appendages; but in *Lamna* I find its fibres woven into those of *M. adductor* to

---

a coalescing of basal parts of rays, refers to the fact that such a coalescing of rays is frequently seen in other parts of the fin, especially anteriorly, he does not see that the basal joints of the rays always are many times longer than the distal, and that this difference of size is also preserved by such concrescences.

[1] The type of the arrangement of the ventral muscular system put down by v. Davidoff (l. c. p. 456) for *Heptanchus* ς, which reminds of the arrangement in *Chimæra*, I have had no occasion to see in any Plagiostome; v. D. asserts to have found it very generally, and refers to *Acanthias* almost as an exception; however, I can with certainty see no other forms mentioned in his text than *Carcharias* as belonging to the same type as *Heptanchus*.

such a degree that it does not appear as an independent muscle, and only artificially is to be separated from the former.

The distal part, the muscular system of the shaft, is typically composed of two muscles: *M. dilatator* (*D*), and *M. compressor* (*sacci*) (*S*).

M. dilatator (*D*) is always very large and powerful; it wraps in a cloak-like manner the appendix-stem until the terminal part, leaving only the lateral surface uncovered, part of which is occupied by *M. compressor*. *M. dilatator* originates forward, either from the appendix-stem only, or frequently also, above the knee of this latter, from the pieces b_1, etc., or from the basale; posteriorly it is attached to the aponeurotic wrapping of the terminal part, or, when covering pieces have been developed from the wrapping, partly to these. Besides fibres of it often go to the skin, and here and there bundles pass into the *M. compressor*. The chief action of this muscle is to bend the terminal pieces together with the soft end-style (ventro-) medially, by which means the terminal part of the appendix-slit is widened; at the same time some of the terminal pieces are often turned from their position of rest in such a manner that they rise through the skin, or are erected so that they stand out free (as the spur or thorn in *Somniosus* and *Lamna*, the claws in *Spinax*, the hook and the spur in *Acanthias*; the large piece T_3 in *Raja*, etc.). When the contraction ceases the appendix-slit will again be narrowed, and the erected skeletal pieces will again be laid, partly mechanically by elastic reaction of the soft connective tissue, but partly also the *M. compressor* will be able to support this latter operation.

Musc. compressor shows in the Plagiostomes so particular a structure, that when it has been examined at all, it has hitherto been misapprehended, the greater part of it being understood as a bag composed of dermal muscles.

This muscle, I suppose, originally occupies in the Plagiostomes a place, similar to that in the Holocephales (see later); i. e. it covers the lateral surface of the appendix-stem, or very frequently only its proximal part, and anteriorly it also reaches on to the piece β and the (two) last rays. Into this muscle, a longitudinal folding of the outer skin penetrates from the dorsal side of the shaft; this folding forms the appendix-slit and the glandular bag, the former leading into the latter. The foremost part of the folding growing on ventrally, carries with it the wrapping muscle, and then both grow on together, and form a singularly thickwalled bag which from the slit-formed opening on the dorsal side grows on between the last ray and the stem skeleton to the ventral side of the fin, where it becomes situated between the outer skin and the ventral ray-muscles. In Sharks the foremost, blind part very often grows much farther forward, not only near to the pelvis, (*Spinax*, *Rhina*, *Somniosus*) but in many, I think in most Sharks it reaches forward of the pelvis (for inst. *Acanthias*, *Scyllium*, *Pristiurus*, *Lamna*, *Selachus*)[1]. and then the bags of the two sides are in contact a long way in the median line (fig. 4). In the Rays the bag is much smaller, (pl. VI, fig. 68), but on the other

---

[1] In *Mustelus laevis* the glandular bag reaches as far forward as to the pectorals; i. e. the part before the pelvis is of far more the double length of that behind it. I have myself only had immature males of *Must. laevis* for examination; but this statement I found on a drawing without any text, left by A. Schneider which, together with other drawings, has been published as an appendix to the fragment left by S.: Studien zur Systematik und zur vergl. Anat. Entwickelungsgeschichte und Histologie der Wirbelthiere (Zool. Beiträge, vol. 2, 1890. The figure in question (pl. 25, fig. 11) is explained as: *Mustelus laevis*. Brustflossen und Bauchflossen mit Saamenblasen. Die Cutis entfernt. Bauchseite.

hand its secreting part is especially developed, and its muscular wall somewhat more complicated. In the part of the *M. compressor* forming the muscular wall of the bag, the direction of the fibres may be rather different, but they chiefly radiate in bent lines towards the periphery, or round this

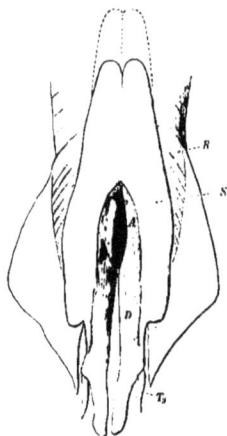

Fig. 4.

Acauthias vulgaris ♂. The ventrals seen from the lower surface. *S M. compressor*, *A M. adductor*, *D M. dilatator*, *R* ray-muscles, T_3 the spur. The stippled contour indicates the anterior extent in another specimen.

to the dorsal surface: this latter is only by loose connective tissue connected with the ray-muscles. In the part in the shaft the direction is more straight, parallel to the axis; this is the case with the fibres covering the lateral surface of the appendix-stem (or a short proximal part of it), as also with those forming the outer, lateral border of the appendix-slit. This lateral part most frequently appears on the dorsal side as an independent muscle, and might be called the „outer lip-muscle“, being, as it were, separated from the other part of the wall of the bag by the attachment of the fin-membrane. By a closer examination and by a transverse section through this region (cp. fig. 1 and 14 in the text) I have been convinced of its forming a whole with the other parts of the muscular wall of the bag, with which also the corresponding part in the Holocephales forms a complete union (see later). A large part of this outer lip-muscle originates anteriorly from the piece *β* and the hindmost ray, or rays; posteriorly it is inserted partly on the inner investment of the ventral marginal cartilage, partly on the aponeurotic covering of the ventral terminal pieces, and acts through this especially on the piece *Ti*. The muscular coat formed by *M. compressor* will by contracting expel the fluid secreted from the epithelium of the bag; but besides its hindmost, lateral part, the „outer lip-muscle“, when it is long and powerfully developed (as in Sharks with a short ventral marginal cartilage, for inst. *Somniosus*, *Spinax*, *Acauthias*, *Rhina*), will act antagonistically to *M. dilatator*, i. e. narrow the dilated terminal part, and lay the erected terminal pieces.

The muscular system of the appendix which here has been briefly represented in its typical characteristics, shows in different Plagiostomes special modifications, as to which the reader is referred to the special part. I shall only here state that the part of *M. compressor* which appears as the „outer lip-muscle“ of the appendix-slit, commonly, as to its size and development, is adjusted to the length of the ventral marginal cartilage; therefore it is very small in *Scyllium* (pl. VI, fig. 66, *S*), and in *Pristiurus*, rather small in *Raja* (fig. 67, *S*); longer and more powerful in *Torpedo*, but especially developed in Sharks as *Somniosus*, *Acauthias*, *Spinax*, *Rhina*, a. o. From the part of *M. compressor* wrapping the bag proper, is in the Rays developed a special muscular layer around the voluminous gland found in these latter. In the Sharks (with the exception of *Rhina*) the inner epithelium of the bag does not form real glands, but only contains secreting cells, and is accordingly very simple as secreting apparatus. In the Rays, however, has been developed a bulky gland protruding as a thick, oval body from the dorsal wall of the bag into its inner space, and almost filling it; when the ventral wall of the bag is opened, this body is immediately seen, and in sound animals it is sometimes seen rather distinctly through the skin[1]. Down the middle of the gland runs straight or obliquely (*Trygon*) a longitudinal furrow, in which is seen a great number of rather large holes with raised margins: they are the excretory openings of collective ducts from a solid mass of large, dichotomously divided, tubular glands. This gland is on all sides until the longitudinal furrow enclosed by a muscular layer, originating from the dorsal muscular wall of the bag. By this special muscular layer the secretion may evidently be ejected into the inner space of the bag, and then by contracting of the muscular wall of the bag itself be driven on, partly through the large opening at the base of the shaft, partly posteriorly through the «tube», formed by the marginal cartilages, and on through the terminal part; in full-grown animals these latter ducts are generally found filled with the secretion. Among the Sharks I have only in *Rhina* found a similar bulky gland, but situated only in the shaft (for further particulars see under *Rhina*).

A survey of the medial fin-muscles in the **females** of the Plagiostomes will show that they are of a considerably simpler structure than those of the male. In the female is found only one single muscle, a *M. adductor pinnæ* (pl. V, fig. 63, 64, *A*) originating in quite the same way as in the male from the pelvis and its aponeurotic prolongation in the ventral median line, and built in a similar manner as to the division of the ventral side in separate bundles, the passing of the foremost lateral part into the ray- muscles, a. s. o.; here, too, the medial marginal portion forms a solid mass, continuing as a posteriorly tapering bundle on to the terminal joint of the fin-stem[2]).

It is then – especially considering the intermingling of fibres that often takes place in the different muscles of the male – an obvious conclusion that an adductor of a similar simple construction as the one, now found in the female, has been the origin of the *M. adductor*, the *M. extensor*, and the *M. dilatator*, perhaps also of the *M. compressor* of the male. When the hindmost joint of the fin-stem developed into the appendix-stem, the distal part of the orginal, simple *M. adductor* might be thought to be brought along at the same time, so that part of the deeper-lying fibres would originate from the stem-skeleton, by which process the *M. dilatator* would arise; while in the proximal part too a group of fibres originating from the stem separated as the *M. extensor* (in *Læmna* this muscle is only part of the *M. adductor*). The *M. compressor* might have the same origin as the *M. dilatator*, but more likely it represents the very hindmost ray-muscles.

In the males of the **Holocephales** (pl. VI, fig. 69–71) the separation between a proximal muscular group and a distal one, placed on the appendix-shaft, is, as before mentioned, more strongly marked than in the Plagiostomes. The proximal group is formed by a *M. adductor*, corresponding to that of the latter, as to the detailed structure of which I refer to the special part; a separate *M. extensor* is not found. The distal part is also here composed of a *M. dilatator* and a *M. compressor*,

---

[1] This gland was already seen long ago. J. Th. Klein (Historiæ piscium naturalis promovendæ missus tertius etc. cum observationibus circa genitales Rajæ maris etc. 1742), as far as I have seen, is the first author, who mentions it. He thinks the gland to be a kind of testis (forte officina seminis), but observes that he has not been able to find any connections with the kidneys, nor with vesiculis seminalibus (adesse tamen possunt). E. Olafsen in his Icelandic voyage II p. 988) takes the same view of the gland as Klein.

[2] The figure of *Acanthias* (given by v. Davidoff, l. c. pl. XXIX, fig. 12, is not correct with regard to the direction of the fibres; so I have given a new figure.

the latter being of special interest with regard to a comparison with the Plagiostomes; it is much thicker than the *M. dilatator*, and covers the lateral surface of the stem-piece b_1, and of the piece b to the terminal part. Into this muscle sinks through the dorsal appendix-slit a continuation of the outer skin as a glandular-bag , which on account of its simplicity might be called rudimentary , when compared to that of the Plagiostomes, as it has evidently remained in a similar stage of development as that, with which it begins in those; by a further development forward and ventrally a quite similar glandular bag would arise as the one described as characteristic in the Plagiostomes. The direction of the fibres of the *M. compressor* is rather peculiar in the Holocephales (see the special part); here I shall only mention that part of the fibres seen dorsally (fig. 70), runs along the lateral edge of the appendix-slit rather straight from the piece β backward in quite the same manner as in the corresponding part, the outer lip-muscle , of the *M. compressor* in the Plagiostomes. The whole structure of this muscle forms, as it seems to me, an incontestable proof as to the correctness of my interpreting the muscular coat of the glandular bag of the Plagiostomes as part of the skeletal muscles proper.

In the female the whole muscular system of the appendix is wanting; according to v. Davidoff the little terminal joint has an attachment for part of the dorsal muscles arising from the wall of the body (i. c. p. 477, pl. XXIX, fig. 18, *As*), corresponding to the attachment of the same muscle on the piece b_1 in the male; just on account of this v. Davidoff explains the terminal joint to be homologous with this piece.

The fin-muscles of the male have been rather slightly treated in the earlier literature; a comparison between several forms has been almost quite out of the question, only a few forms having been described. Thus among the Sharks *Acanthias* has already been mentioned by Bloch, among the Rays some *Raja*-species by several authors (*Raja radiata* very briefly and incompletely by Bloch, *Raja circularis* [or *clavata*] by Duvernoy, *R. clavata* by Vogt & Pappenheim and later by Moreau), *Chimæra monstrosa* by v. Davidoff. Petri alone has examined several different forms and tried to make a comparison, but he cannot be said always to have been successful or to have found the correct interpretation. While he upon the whole pretty correctly has interpreted the muscle I have called *M. adductor*, his *M. flexor pinnæ*, or *pterygopodii*, a name rejected by me as presumably not suitable, and *M. dilatator*, a name introduced by him (at all events in *Scyllium*, *Acanthias* and *Torpedo*), the other muscles have either been misapprehended or not at all mentioned. The *M. extensor* he has only seen in *Scyllium* and *Raja*, where he calls it *M. flexor pterygopodii interior*, and of my *M. compressor* he has only mentioned the part, which I have called the outer lip-muscle (of the appendix-slit), in *Acanthias* and *Raja*, and with different appellations, respectively as *M. levator* (of the spur) and as *M. flexor biceps* (which latter name is also given to a quite different muscle in *Scyllium*), and he has assigned to it different, partly misapprehended, functions. It has already been observed that both Petri and all other authors, who have mentioned the glandular bag , have understood the muscular wall to be a separately developed dermal muscular system, and consequently omit it by the mentioning of the fin-muscles proper. In the special part account will be rendered of the earlier literature, and the particular works will be referred to.

2. Special Part.

Selachoidei.

Spinacidæ.

Acanthias vulgaris Risso.

(Pl. 1, fig. 10, 11.)

The common picked Dog-fish has been so often examined that I think a more particular description of the external features of the copulatory appendages to be superfluous; I may refer to Petri[1] (with regard to whose description, however, I must remark that the investment with dermal teeth at the places of transition to naked parts does not cease gradually, but is quite sharply bounded; the dorsal side is wholly naked, as is also on the ventral side the hindmost point of the terminal part), as also to the earlier description by Bloch[2] and Home[3]. In a specimen of the length of 64^{cm} the following measures were found:

Length of the appendix (from the fore-edge of the cloaca) . . 6,5^{cm}
—   - - part free of the fin . . . . 3,1^{cm}
--   - - terminal part . . . . 2,2^{cm}
- - appendix-slit . . . . 4,2^{cm}
Breadth of the appendix . . . . ab. 1^{cm}

The skeleton has not been quite correctly described by any of the earlier authors[4].

Between the basale and the appendix is found only one short joint (b_1), and besides the dorsal piece (β^5); this latter articulates anteriorly with the basale, posteriorly with the appendix-stem b, and medially with b_1; its lateral edge is convex, projecting somewhat in the shape of a roof over the two hindmost rays; these rays are borne by the piece b_1, and are often coalesced; they are stronger and longer than the last ray but two, which latter comes from the basale.

The stem of the chief piece of the appendix has a length like $B - b_1$, and proximally towards its articulation with b_1 is found a ridge (at b in fig. 10) projecting in a somewhat keel-like manner; in the hindmost half it has laterally a little trough-like hollow. The soft end-style is short[6], flatly rounded, and reaches not nearly to the end of the terminal part. The dorsal marginal cartilage[7] (Kd) can forward be indistinctly traced as a rounded ridge to about the letter x in fig. 11 (it is more

---

[1] l. c. p. 300, pl. XVII, fig. 5, A.
[2] l. c. 1788, p. 9, pl. 2, fig. 1.
[3] On the Mode of breeding of the Ovoviviparous Shark etc. Phil. Trans. 1810, Pt. II, p. 205, pl. IX and X; in the lastmentioned place the ventrals and the appendages have been drawn in a position, which they scarcely naturally would be able to have.
[4] Drawings are found not only in Bloch, Gegenbaur and Petri, but also in Molin: Sullo scheletro degli Squali, pl. III, fig. 7; Memorie dell' Ist. Veneto, vol. 8, 1859, but without any explanation or description in the text.
[5] Gegenbaur, fig. 16, b; Petri, fig. 5 D, r'.
[6] Gegenbaur, fig. 17, i; it has been quite overlooked by Bloch and Petri.
[7] Mentioned neither by Gegenbaur nor Petri, The hindmost end of it is the Processus a am Schienbein (l. c. fig. y of Bloch. Neither of these authors have seen independent marginal cartilages in *Acanthias*.

distinct, when the piece is dried); posteriorly it is distinctly elevated as an edge of the appendix-slit. The ventral marginal cartilage (*Kv*) is shorter, resembles the corresponding one in the Greenland Shark, and has, as in the latter, a plate-like part[1]) folded to the dorsal side; on the concave inner side it has furthermore a strong, elevated process; in the furrow between this process and the folded part the proximal end of the thorn is placed.

There are four terminal pieces.

Td[2]) is narrow, with the foremost part of its medial edge closely connected with the end-style, and behind this with the edge of the ventral piece *Tv*; distally it takes the form of a flattened, sharp-edged hook; this hook-shaped part rises uncovered through the skin, is smooth, shining, and dentine-like. *Td* is with part of its lateral edge connected with a quite thin, plate-formed piece[3]), *Td*, also anteriorly connected with the marginal cartilage *Kd*; it is placed in the skin forming the dorsal lip of the appendix-slit of the terminal part, and corresponds to the piece *Td₁*, indicated in the Greenland Shark.

The ventral terminal piece, *Tv*[4]), is considerably broader and longer than the dorsal one, rounded on the ventral (outer) surface, hollowed like a spoon towards the appendix-slit; except the hindmost part it is firmly calcified; the foremost part of the medial edge is connected with the end-style, and behind this with *Td*, the hook of the latter lying freely in the outermost spoon-like end of the former piece; in the proximal end it has medially an articular process for articulation with the above mentioned process of the concave side of the marginal cartilage *Kv*, and its lateral edge is firmly connected with a strong, thin membrane (fig. 11, *Tv₃*), serving in the foremost part for attaching the outer lip-muscle of the glandular bag; this membrane then corresponds to the similar, but thicker one in the Greenland Shark, and to the piece *Tv₂* in *Spinax*.

The fourth terminal piece, *T₃*, is the one called the "spur"[5]) by the different authors; with the proximal, somewhat head-shaped end it is attached inside of the folded plate of the marginal cartilage *Kv* to the above mentioned process, and to the proximal and lateral end of the piece *Tv*; it is formed as a triangular thorn or spine, longitudinally somewhat twisted, with two concave surfaces; it is firm, shining, dentine-like, and the greater part of it is uncovered by the skin. It can be moved quite in the same manner as the corresponding spine in the Greenland Shark.

The muscular system. The *M. adductor* shows the general typical relations. The *M. extensor* reminds very much of the same one in the Greenland Shark; as in the latter it has here its origin on the medial side of the basale and *b₀*, stretches over the knee of the appendix-stem as a thin, flat covering over the *M. dilatator*, and inserts itself along the boundary line of the dorsal marginal cartilage.

The *M. dilatator* originates proximally with a dorsal portion at the same place as the *M. extensor* and quite covered by it, that is to say some way up on the basale; on the ventral side its proxi-

[1] Bloch, Processus *d*; Gegenbaur, fig. 15, 16, *a*; Petri, fig. 5, *D, E, pr*: regarded by all only as a process on the chief piece.

[2] Bloch, der Haken, fig. 2, *e*, fig. 6; Gegenbaur fig. 16, 17, *o*; Petri, fig. 5, *hh*.

[3] Petri, fig. 5, *la*; it is neither mentioned nor drawn by Bloch or Gegenbaur.

[4] Bloch: der breite Knochen, fig. 2, *d*, fig. 5; Gegenbaur, fig. 15—17, *e*; Petri, fig. 5, *b'''*, he interpreting it as the terminal joint of the stem.

[5] Bloch, der Sporn, fig. 2, *c*, fig. 4; Gegenbaur, fig. 15, 16, *a'*; Petri, fig. 5, *sp* and *ca*.

mal origin accordingly is much more backward, at the distal end of the *M. adductor*; it is inserted as usual, its aponeurosis being especially attached to *Ti* and *Td*; the latter piece, the hook , is turned (round its medial edge as the axis) out of its position in the spoon-shaped end of the former, when the muscle is contracted during the dilation.

The part of the *M. compressor* wrapping the bag, is much distended, and consequently rather thin, corresponding to the considerable extent of the bag anteriorly (see fig. 4 in the text). The part inserted on the lateral surface of the appendix-stem, is very small, reduced to a few bundles of fibres on the proximal end of this part of the skeleton, which otherwise is almost quite enclosed by the *M. dilatator*. The part, which as outer lip-muscle forms the lateral limit of the appendix-slit, seems to me to receive in its surface some fibres coming from the muscular layer originating from the lateral muscles of the body, but otherwise it originates as usual on the hindmost rays and on *β*; it is inserted with a kind of tendon in the above-mentioned membrane on *Tv*, and consequently it acts antagonistically against the *M. dilatator*, and at the same time lays the spur *T₃¹⁾*.

Spinax niger Bonap.

(Pl. 1, fig. 12, 13.)

The very peculiar-looking appendages in this common Shark have singularly enough been very little mentioned by earlier authors, and by many, also among the later, they are not mentioned at all. Gunnerus[2], in his description of the Sort-Haa , says: they (i. e. the two *Membra genitalia*) were supplied with some sharp bony spines, such as I have seen on the *Membra* of several Rays, when the ends have been turned inside out. Kroyer[3] says: At the end of the copulatory appendages of the males are found three crooked thorns or horny claws, and a tapering dermal flap, which behind projects a little over these claws. The claws are movable against each other, and form a kind of prehensile organ. In the position of rest they are hidden between a pair of small cartilaginous plates, and the skin covering these plates. This is the most complete, and also, I think, the most correct description I have seen[4]. Duméril[5] gives a drawing of the appendix, but with no explanation whatever (nor in the text neither); the drawing is rather difficult to understand, neither is it correct; thus the dermal flap mentioned by Kroyer appears in this figure as a thorn, although it is

---

[1] Petri (l. c.) designates this part of my *M. compressor* as *M. levator* (fig. 5, *mt*), and attributes to it a dilating effect, having allein die Aufgabe diesen (den Sporn) zu heben , and thus he in this place speaks of two dilating muscles. The incorrectness of this, however, is easily pointed out. Contrary to Petri, Bloch upon the whole has a correct understanding of the mobility of the spur, speaking (l. c. p. 13) of einen sehr sonderbaren *Mechanismus*. Davon mir wenigstens in der Anatomie kein ähnlicher bekandt ist . Bloch has a chiefly correct description of the muscular system; he distinguishes between three muscular portions, the first of which being the ventral ray-muscles, the second, which he compares to the adductor femoris in man, is my *M. adductor*, the third *M. dilatator* + my *M. extensor*. He describes the glandular bag as a particular organ, to which he does not ascribe any muscular walls, as he supposes that the other (2) muscles expel its klebrigte Feuchtigkeit . Neither has Petri seen my *M. extensor* as a separate muscle in *Acanthias* (see his fig. 5, *B*, and the description p. 302); but it is also to be acknowledged that in this species it is very closely connected with the *M. dilatator*, especially proximally.

[2] Throndhjemske Selskabs Skrifter II, 1763, p. 319.

[3] Danmarks Fiske vol. III, 1852—53, p. 908.

[4] Müller & Henle, System. Beschr. der Plagiostomen. 1841, p. 86, say: Kein Dorn an den männlichen Anhängen ; founded, I suppose, on a position in which only the soft dermal flap is seen.

[5] Hist. nat. des Poissons, vol. I, 1865, the atlas, pl. IV, fig. 13.

quite soft. Lilljeborg[1] only says: The copulatory organs of the male are small and pointed, and reach only a little behind the ends of the ventrals; they are until towards the end coalesced with the ventrals.» As no thorns are mentioned, L. must have examined only undeveloped appendages.

The appendix, when fully developed, is short, clumsy, thick, and reaches only a very little farther backward than the end of the fin-membrane, the free part of which is also very short. Dermal teeth are not found, neither on the dorsal, on the medial, nor on the greater part of the ventral side, except on this latter laterally, near the fin-membrane. In the numerous, developed appendices, examined by me, the terminal part was always very much dilated, and such was also the case in the specimens, I have caught alive; in the dilated state the terminal part stands almost at a right angle to the stem, its hinder end with the soft dermal flap (a) pointing inward towards the middle line; the dilated part of the furrow then looks like a concave sole of the foot, in whose «heel» is seen the opening, through which the secretion of the glandular bag is probably ejected. Three polished, hard points protrude like claws through the skin, one at the dorsal lip of the furrow, the second at the ventral lip, and the third, and longest, juts out, ventrally and laterally, from the spot, where the fin-membrane becomes free of the appendix.

Fig. 5.

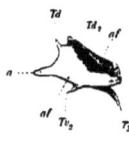

Fig. 6.

Fig. 5. *Spinax niger*. The appendage of the right side with part of the fin-membrane, seen from the dorsal side, somewhat enlarged. The terminal part is dilated. *f* the folded, free end of the fin-membrane: at *o* the fin has been cut from the body. *a/*, the dilated part of the appendix-slit.

Fig. 6. The dilated terminal part, seen from behind. *a* the soft terminal flap. *af* the spot where the appendix-slit passes into the dilated, terminal part of the furrow.

In specimens of the length of 35.5^{cm}—38.5^{cm} the following measures are found[2].

Length of appendix (from the fore edge of the cloaca) . . abt. 2.5^{cm}--3.5^{cm}

—   - the part, free of the fin . . . . •   1.2^{cm}

    - the terminal part . . . . •   1^{cm}

    - the appendix-slit . . . . -   1.8^{cm}

Breadth of the appendix . . . . •   0.6^{cm}—0.8^{cm}

The skeleton. Between the basale and the appendix are found two small pieces (b_1 and b_2), each bearing one of the two hindmost rays (accordingly $b_1 + b_2$ in *Spinax* = b_1 in *Acanthias*); nevertheless these rays may be found coalesced, and are, as usually, directed straight backward, parallel to the appendix. The piece *i* is relatively somewhat longer than in *Acanthias*, but of a similar form.

The axial part of the chief piece of the appendix is somewhat more clumsy than in *Acanthias*, but otherwise of a similar form, and also supplied with a short, soft end-style; including this latter the stem is only a little longer than the basale. The marginal cartilages, too, show chiefly the same relations as in *Acanthias*.

[1] Sveriges och Norges Fiskar. vol. 3. 1891, p. 677.
[2] It is somewhat difficult to obtain exact measurings on account of the terminal part being bent.

The terminal pieces are 5.

The dorsal one, Td, is somewhat s-shaped, round, and articulates medially with the end-style, while the hindmost part of it projects through the skin as a curved, polished claw; as in *Acanthias*, it is united with a thin lamellar piece, Td_i, which piece, with the exception of the hindmost point, is quite covered by the skin forming the dorsal lip of the furrow.

The ventral piece Tv is also somewhat s-shaped, broader than the dorsal one, thick at the base, becoming thinner distally and laterally; it is concave like a spoon on the side towards the furrow, on the other side rounded. At the proximal part of the lateral edge it is firmly united with a hard, dentine-like piece Tv_z, which in *Acanthias* is only represented by an uncalcified membrane. This piece is before (proximally) prolonged to a long, flat end, behind (distally) to a shorter one, projecting through the skin as the before mentioned claw in the ventral lip of the furrow; the piece is rather narrow, ventrally concave, dorsally rounded. In moving it follows the piece Tv.

The last piece T_3 corresponds to the thorn in *Acanthias* and *Somniosus*, and is also here formed as an elegant, bent, rounded and completely smooth thorn with the proximal end head-shaped.

It is quite out of the question that these claws, as supposed by Kroyer, should be able to act as a prehensile organ, as they cannot properly be moved against each other; but they will be very able to fix the appendix firmly in a hollow, as by the dilatation of the terminal part their points are turned in three opposite directions, as may be seen from fig. 6 in the text.

The muscular system. From the *M. adductor* has been separated a long, flat bundle as a particular muscle originating before from the medial aponeurotic stripe together with the other fibres of the *M. adductor*, and then on the dorsal side passing obliquely over the *M. extensor* and next over the *M. dilatator*; on the appendix it follows the appendix-slit, and forms together with the *M. dilatator* the medial lip of this slit; partly it is attached in the skin of this lip, but chiefly on the proximal end of the piece Td_z. This muscle evidently is instrumental in increasing the dilation of the terminal part, which dilation, as has already been indicated, seems to be especially great in *Spinax*.

The *M. extensor* is almost as in *Acanthias*, that is, not sharply bounded from the dorsal part of the *M. dilatator*.

This latter, on the contrary, is on the ventral side distinctly bounded from the *M. adductor* by a line running obliquely from the lateral side down towards the medial side. Its aponeurosis, as in *Acanthias*, is especially attached to Td and Tv.

The glandular bag (the *M. compressor*) does not in any of my numerous specimens reach quite to the pelvis, and accordingly it must be termed proportionally small. Its outer lip-muscle as usual originates from the piece β and the hindmost rays, and is with its principal portion very distinctly inserted on the piece Tv_z, with another portion on the folded part of the ventral marginal cartilage (not on the thorn T_3).

Scymnus lichia Bonap.

A skeleton in the Zoological Museum (from V. Frič in Prague).

In this specimen the appendix only reaches a trifle farther backward than the fin-membrane, and the condition of the terminal skeleton makes it probable that the organ is not fully developed.

Between the basale and the appendix-stem is found one piece b_i bearing the two hindmost rays. The piece β is rather large, flattened, with an edge turned towards the dorsal side.

The appendix-stem is as long as $B + b_i$; its proximal part below the knee is somewhat bent, medially convex, otherwise of a similar form as in the Greenland Shark, i. e. distally lanceolate; the end-style is very short. The dorsal marginal cartilage is a very narrow ridge, reaching forward almost to β; the ventral one is much longer than in the Greenland Shark, occupying almost the whole length of the appendix-stem as a rather high, firm, and hard lamella, the distal part of which forms a but small, very narrow, folded plate, properly speaking only an indication of such a one.

Among the terminal pieces the piece Td is still quite soft, not separated from the other tissue; Tv on the contrary is hard, and reminds, as to its form, of the corresponding piece in the Greenland Shark. T_3 is present, but small, and no doubt not yet quite formed; whether in the developed organ it is hidden by the soft tissue — so that the observation by Müller & Henle: Die männlichen Anhänge ohne Stachel (d. c. p. 91) so far may be justified — I must leave undecided; the observations of these authors concerning the ventral appendages are however, as it turns out, often quite unreliable.

Scylliidæ.

Scyllium canicula (L.).

(Pl. II, fig. 16, 17).

The copulatory organs are mentioned by several authors, generally, however, without any particular description, as these authors especially attach importance to one peculiarity in the male, which (in all stages) forms an easy distinctive mark between *Scyllium canicula* and *Sc. stellare* (*catulus*)[1], viz. that the ventrals are completely coalesced dorsally of the appendages, and in the middle of the hindmost edge of this coalesced part only a small incision is found. By a fold of the fin-membrane, passing over the proximal part of the appendages, these are also partially covered on both sides ventrally, and thus they are placed as tongues in a bell, which is open on the lower side, their hindmost ends reaching to or even farther (abt. 5^{mm}) than the hindmost edge of the bell[2]. The whole dorsal side (i. e. the side towards the body) of the coalesced ventrals is covered with dermal teeth and pigmented (spotted like the skin of the animal in other places), and this covering is continued round the edge to the ventral side, where it is quite sharply limited; the other ventral part of the coalesced fins (the part in contact with the dorsal side of the appendages) is naked, unpigmented, and soft.

The appendix (in two specimens, when measured from the cloaca, abt. 43^{mm} long, abt. 6^{mm} broad at the base of the terminal part, which is of a length of abt. 24^{mm}) is straight, posteriorly some-

[1] See for inst. Müller & Henle, l. c. p. 7, 10. Kroyer, l. c. p. 824. Duméril, l. c. p. 316, 317. Lilljeborg, l. c. p. 650. Petri. l. c. p. 303, and fig. 6.

[2] The words of Lilljeborg l. c. p. 650: The male has small copulatory organs, not reaching to the hindmost points of the ventrals, and scarcely of half of the above given length of these fins do not apply to the developed state. Neither can the figure 6 of Petri represent the developed appendages, and it is upon the whole bad; the appendages are in this species never so clumsy; the description at p. 303 is only ill adapted to *Sc. canicula*, and not very well to *Sc. catulus*.

what conically tapering, on the greater part of the surface covered with dermal teeth; only immediately at the cloaca the dorsal side is naked, as is also the outermost point of the appendix, which is soft and papillous; from here a naked, depressed stripe reaches forward on the medial side of the terminal part [1]. On this part the dermal teeth have another shape than elsewhere on the animal, being longer and more pointed, like small thorns with the points turned t o w a r d s t h e b a s e of the appendix; accordingly the hinder part of this is rough to the feeling when rubbed b a c k w a r d, contrary to what is the case elsewhere on the animal. The appendix-slit is covered in the terminal part by a thin, soft membrane arising from the dorsal (inner) lip; when this membrane is thrown back, the furrow is found to be open as usual; but above the terminal part it is only represented by a groove in the skin, not very deep; the slit, which in the Sharks, hitherto mentioned, is quite open, is in this animal under the dermal furrow by coalescing formed into a tube reaching to the base of the organ near the cloaca, and first here an opening is again found, an oval aperture through which a sound may be brought into the glandular bag. This latter accordingly has two outlets, one at the base of the appendix, the other between the movable parts of the terminal part [2].

T h e s k e l e t o n. Between the basale and the appendix is found one very small piece (b_1) bearing no rays; the piece β is also inconspicuous, somewhat triangular, with a broad articulation before with the basale, a narrow one behind with the appendix-stem.

The appendix-stem is of about the same length as the basale; it is calcified to a rather considerable degree; the soft end-style reaches to somewhat more than half the length of the terminal part.

Both marginal cartilages are specially strongly and peculiarly developed, which will be seen from fig. 16 clearer than from a description. The dorsal one (Rd) reaches (as is usual) somewhat further forward than the ventral one, but in the dorsal middle line it joins with the latter for a long way by a firm ‹suture›, so that the two cartilages together with the stem form a complete, firm tube, open before where the glandular bag joins it, and behind at the terminal part. Thus the part of the ventral marginal cartilage assisting in the forming of this tube, corresponds to the folded plate of the ventral marginal cartilage in the before mentioned Sharks [3].

The number of t e r m i n a l p i e c e s is four [4], completely corresponding to those in *Acanthias*. Td is narrow, somewhat triangular; along the side towards the furrow it is connected with a thin, style-shaped piece, Td_2 which proximally becomes broader, and reaches a little under the dorsal marginal cartilage. Tv is broader, lengthened-oval, rounded on the outer side, towards the furrow slightly hollow, thick, and solid. Between its proximal end and the ventral marginal cartilage is inserted a well developed piece, T_3, which is not formed as a thorn, nor can it be erected to such a position, as

---

[1] At *x* in the fig. 6 of Petri.

[2] D a v y, J.: On the Male Organs of some Cartilaginous Fishes, Phil. Tr. vol. 10, 1839, p. 146, has already mentioned this fact in *Scyllium Edwardsii*; Petri represents it l. c. p. 304.

[3] As Petri has not seen the marginal cartilages as such in *Acanthias*, he has in *Scyllium* understood them to be something particular in this genus.

[4] When Petri also finds four pieces in *Scyllium* it arises from his counting the end-style of the stem (b''' fig. 7, *C*); he has really overlooked one piece, viz. Td_2.

the corresponding piece in the hitherto mentioned Sharks. All these terminal pieces are hard, white, china-like, but none of them protrudes with any part through the skin.

The muscular system is as in *Sc. stellare*, where it will be more particularly mentioned.

Scyllium stellare (L.)

(Pl. II, fig. 18—19; pl. VI. fig. 65. 66.)

As upon the whole the ventrals of the male as to contour and shape are different from those in the preceding species, so it is also the case with the appendices. The ventrals are also here coalesced[1], but only for a short way (in one specimen of the total length of 90^{cm} the coalesced part has a length of 16^{mm}); the small cut in the posterior edge in *Sc. canicula* has here become a large slit (in the specimen mentioned above about 26^{mm}); the dermal teeth also spread to a greater extent on the ventral side of this part of the fin. As furthermore no lateral fold of the skin is found on the ventral side covering the base of the appendices, and corresponding to the one mentioned in *Sc. canicula*, no bell is formed here.

The appendix reaches just outside the posterior fin edge; it is far more big and clumsy than in *Sc. canicula*, but still the details remind of the latter, they are only coarser and more conspicuous.

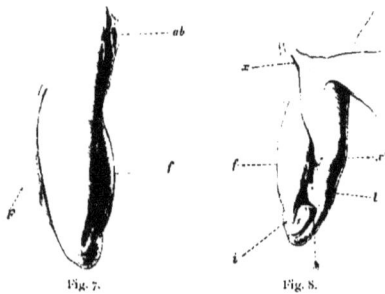

Fig. 7. *Scyllium stellare*. The appendage of the right side seen from the ventral side; about the natural size. *ab* abdominal pore. *F* fin-membrane, *f* winglike process.

Fig. 8. The same appendage seen from the dorsal side; the coalesced part of the membrane of the ventral is cut up, and thrown back. The arrow indicates the direction, in which a sound may be brought into the appendix-canal.

Fig. 7.   Fig. 8.

In a specimen of a total length of 90^{cm} the following measures were found:

Length of the appendix from the fore-edge of the cloaca to the hindmost point . 61^{mm}

— · · free part. . . . . 36^{mm}

— · · terminal part . . . . 34^{mm}

Breadth of the appendix above the terminal part . . . . 14^{mm}

— · · across - — · . . . . 16^{mm}

The terminal part is relatively larger than in the preceding species, and its peculiar appearance is especially caused by the strongly developed process *f*, which is only indicated in the preceding species. This process is on the ventral side (fig. 7) hollow, and the bottom of this hollow is naked, which nakedness continues on the soft, outermost point. The greater part of the appendix is also

[1] When Lilljeborg l. c. p. 655 tells that the ventrals in *Sc. stellare* are not coalesced, he is not quite right. Müller & Henle l. c. p. 10 state the fact correctly.

here covered with dermal teeth; besides the parts mentioned only the surroundings of the anterior aperture of the glandular bag are naked. The points of the dermal teeth are also turned towards the base of the appendix; they are longest and most pointed on the dorsal side of f and l. The appendix-slit is closed (to an extent of abt. 15^{mm}) in advance of the terminal part, as may be seen by throwing back the dermal lip x x' in fig. 8; accordingly we have as in *Sc. canicula* two outlets for the secretion of the glandular bag.

The skeleton in its main features is as in *Sc. canicula*, but the appendix-part of it is much more clumsy and peculiarly twisted. One small b_1 without rays, and a little β with rounded contour are found[1].

The appendix-stem, from the articulation with b_1 to the end of the style, is of the same length as the basale; it is somewhat bent with medial concavity; the end-style of about half the length of the calcified stempiece; at the distal end of the former the medial edges of both the adjoining terminal pieces form a rather sharp knee.

The marginal cartilages are principally like those in *Sc. canicula*; Rd is posteriorly somewhat longer than Rv, and is distally and medially a little hollow.

The terminal pieces are four, three of them white and hard. Td is formed somewhat like a roof and as broad medially as Tv is ventrally; Td_2 is mainly as in *canicula*; Tv is rounded on the outer side, somewhat concave towards the slit, T_3 in my specimen is not calcified; but a soft, fibrous cartilage, joining with Tv and placed in the lip l, in my opinion represents this piece[2]. As in *Sc. canicula* none of the terminal pieces are seen through the skin.

The muscular system. From the medial marginal part of the *M. adductor* have been branched off two separate muscles: fig. 65, fig. 66 a_1 and a_2.

If we look at the ventral side (fig. 65) the fibres of the marginal part are seen as a powerful muscle a_1, anteriorly originating from the medial aponeurotic stripe, and posteriorly inserted on the proximal part of the appendix-stem close to the ventro-lateral edge of the skeletal orifice for the glandular bag; but part of its fibres attaches to the basale, and another part runs into the *M. dilatator*. Looking at the dorsal side (fig. 66) we find the edge formed by another muscle a_2, anteriorly only indistinctly separated from a_1, but posteriorly distinctly enough, as here a foremost portion of the *M. dilatator* originating from the medial side of the basale, wedges in between both. This muscle a_2 distally joins with the *M. extensor* (E), and together with this is inserted by a tendon below the knee of the appendix-stem.

The *M. dilatator* is enormously thick, and originates with the greater part of its mass from the appendix-stem until the boundary of the marginal cartilages, but, as already mentioned, a portion of it arises from the medial side of the basale; part of this muscle distally joins in the composition of the peculiar process f (it is the same in *Sc. canicula*, where this process is much less conspicuous), which by no means, as Petri says, is composed exclusively of »verfilztem Bindegewebe«.

---

[1] Petri l. c. fig. 7 C has distally of β (r' in Petri) another little piece (r''), which is not found at all in my specimen, and which upon the whole I do not think to be normal (originating from a rupture?); furthermore a piece (*mr*) which he (p. 305) compares to a »knee-cap«; this is, however, scarcely to be regarded as a particular piece, but, I suppose, only a strongly calcified eminence on the stem.

[2] Petri, fig. 7 C. *x*.

M. compressor. The bag-formed part of this muscle is rather long, and reaches considerably forward of the pelvis. On the contrary, the part forming the -outer lip-muscle is rather small (smaller than in my figures); as usual it originates from the stem-skeleton (*d*) and from the hindmost rays, and is inserted on the proximal edge of the ventral marginal cartilage; it will here scarcely be able directly to contract the dilated terminal part[1]).

Pristiurus melanostomus (Bonap.).

(Pl. II, fig. 20, 21.)

The ventrals of the male of this species are also dorsally coalesced in a similar manner as in the preceding two Scylliidæ[2]), but to a still less extent than in *Sc. stellare*, and a deep curve separates the coalesced part into two fin-laps. The appendices reach far behind the fin-membrane[3]), in a specimen of the length of 78^{mm} to 23^{mm} behind the point of the fin-membrane; the whole length, from the hindmost edge of the cloaca, is 50^{mm}; the part quite free of the fin is 35^{mm} long; the largest breadth of the organ is about 7^{mm}; the terminal part has a length of about 25^{mm}. The ventral side is covered with dermal teeth, except the hindmost, soft, as it were, convoluted part (abt. 9^{mm} long), on which still scattered groups of teeth may be seen; the dorsal side is naked, as are also the adjoining parts of the medial side, where they are covered by the fin-membrane; on the free edge of the lip *l* a few scattered rows of dermal teeth are seen. The dermal teeth are generally very fine; as in the foregoing species their points are on the terminal part turned towards the base of the appendix. The dentition on the coalesced fin-parts is as in *Sc. stellare*.

The peculiar appearance of the appendix will be seen with sufficient distinctness from fig. 9. The furrow anteriorly is opened by a large, easily distended slit of a length of 8—10^{mm}; behind this slit it is closed for an equal length, and again open in the terminal part. In spite of the great dissimilarity in general when compared with the appendix of the preceding Scylliidæ, a closer examination will show a rather considerable similarity with these, especially with *Sc. stellare*: corresponding to the peculiar process *f* of the Scyllia is found a thin, soft dermal process, which may be folded towards the furrow (as in fig. 9), or spread in a wing-like

Fig. 9.

Pristiurus melanostomus. The right appendage seen from the dorsal side; about the natural size. The fin-membrane is cut through, and thrown back. *o* the larger basal opening of the appendix-slit; between the asterisks it is closed by coalescing. The other signs as in fig. 8.

[1] This part of the *M. compressor* has been quite overlooked by Petri, who has seen and drawn the other muscles, and given them the following names (see l. c. pl. XVII, fig. 7, A and B):
The muscle here marked *A* (the chief portion of the *M. adductor*) = *fl. m. p.* i. e. *flexor major pinnæ.*
    •   —   —   *a₁* = *fl. p. b.* i. e. *flexor pterygopodii biceps.*
    •   —   —   *a₂* = *fl. p. ex* i. e. *flexor pterygopodii exterior.*
The *M. extensor* here marked   *E* = *fl. p. i.* i. e. *flexor pterygopodii interior.*
What Petri calls flexion must, I think, rather be regarded as an abduction connected with an extension of the appendix.
[2] The expression used by Lilljeborg l. c. p. 660 their inner edges are not coalesced is accordingly not quite correct. The appendices are shortly described at p. 662.
[3] Comp. Gunnerus: Om Haae-Gielen, pl. I, *f.* (Trondhjemske Selsk. Skr., II.)

shape to the medial side; to the lip *l* which in the Scyllia is turned into the furrow, corresponds the part in *Pristiurus* marked with the same letter, to the naked dermal fold *i* in one corresponds the naked dermal fold *i* in the other, etc.

Also the skeleton shows the near relation to the other Scylliidæ. Between the basale and the appendix is found (ventrally) a very small, quite rudimentary piece b_i which of course bears no ray; dorsally is found an also very small piece *j*.

The appendix-stem and the marginal cartilages are much like those in the Scyllia. The stem is twisted longitudinally in a similar manner as in *Sc. stellare*, but is not bent medially. The dorsal connection of the marginal cartilages, however, is not so close as in the Scyllia; the two pieces may here be forced a little from each other.

The number of terminal pieces is five, if to the terminal pieces we will count a piece, Rd_i, which has not been found in any of the Sharks, mentioned in the foregoing; it is joined movably to the hindmost edge of the dorsal marginal cartilage, and is situated in the dermal fold below the asterisk in fig. 9.

Td and *Tv* are long and narrow, and form at the end of the style a similar kneeshaped curve as in *Scyllium stellare* (in *Sc. canicula* it is only indicated); a slightly calcified or almost quite soft piece Td_i is found, projecting forward under the edge of *Rd*; *Tv* proximally forms a rather broad plate, to the dorsal edge of which is attached a leaf-shaped, somewhat bent piece T_3. All the pieces are completely hidden in the skin.

The muscular system is substantially quite the same as in the Scyllia, the only difference being that the outer lip-muscle seems to be still less developed in *Pristiurus*.

Lamnidæ.

Lamna cornubica (Gmelin).

(Pl. II. fig. 22, 23.)

In a specimen of the length of $2^m\ 5^{cm}$, which in the beginning of November 1897 was driven on shore on the western coast of Jutland, the appendix has a length of 21^{cm} [1]) and a largest breadth of 4^{cm}; the terminal part is 7.5^{cm} long. The whole ventral surface is densely covered with dermal teeth quite to the end; this investment ceases with a strongly marked boundary line on the medial surface, which is quite naked to the terminal part; this latter being almost quite covered with teeth until the margins of the appendix-slit, also on the dorsal side; the other parts of the flat dorsal side of the shaft are naked, and these naked parts are laterally marked off from those covered with teeth by a rather deep longitudinal dermal fold. Apparently the appendix-slit from before the terminal part and to a larger foremost opening at the base of the appendix is closed as in the *Scylliidæ*; but in reality it is open, and for the whole way it is possible, though with difficulty, to press a sufficiently thin sound in between the margins of the marginal cartilages. On the medial side, immediately be-

[1] Lilljeborg l. c. p. 625 gives for a specimen of the length of 2.4^m a length of 25^{cm} for the appendix.

fore the terminal part, is seen a small opening (l in the fig. 10 in the text) of a length of $1^{1/2}$^{cm}; this opening leads into a deep, pocketlike invagination of the skin, lined with a soft continuation of this, similar to a mucous membrane. On the dorsal side of the terminal part, at the lateral base, a polished skeletal piece projects uncovered by the skin and like a thorn (T_3). The terminal part is easily bent ventrally; if bent in that way, the thorn, as in many other Sharks, will rise mechanically, and stand out horizontally; it will immediately lie down again, when the terminal part is let loose. A dermal fold supported by the skeletal piece Td_1 is prolonged forward into the tube formed by the marginal cartilages in such a manner, that this tube gets two outlets, one on each side of the lamella concerned; but the appendix-slit proper is situated laterally of this piece (af in fig. 10).

The skeleton. Between the basale and the skeleton of the appendix is found (almost as in the Scylliidae) one very small piece b_1, highest on the medial surface, and otherwise quite low, that is to say, wedge-shaped; in connection with the distal end of the basale and the proximal end of the appendix-stem it bears the hindmost ray, which at the base is rather broad. The piece β is pretty well developed, and, as is usual in Sharks, connects the basale with the appendix-stem.

The appendix-stem is very long, twice as long as the basale $+ b_1$; proximally it is only a little calcified (comp. *Selachus*), but else it is firmly calcified in the surface until the terminal part, where it forms a very long style, reaching to the hindmost end of the terminal part; this style for the hindmost two third parts is calcified in the surface; its soft basal part is situated immediately under the above mentioned «pocket»; the distal end of the *Musculus dilatator* passes into the firm, fibrous ventral wall of this pocket, in such a manner that its aponeurosis is firmly inserted in the perichondrium above the calcified part of the style, as well as in the corresponding places of the two adjoining calcified terminal pieces Td and Tv; the soft part of the style and the joints between the marginal cartilages and the two skeletal parts Td and Tv will then act as a kind of articulation [1]).

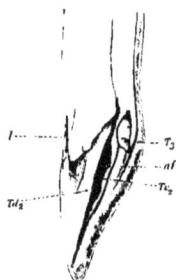

Fig. 10.

Lamna cornubica. The hindmost part of the right appendix seen from the dorsal side; considerably reduced. l the opening of a pocketlike invagination of the skin. Tv_2 a dermal fold containing no skeletal piece.

The marginal cartilages are very long, hard, and thick; forward they reach almost to the beginning of the appendix-stem; the dorsal one reaches somewhat longer forward and also somewhat further backward than the ventral one. In the greater part of their length the two cartilages are in contact with their margins; proximally the dorsal one is covered a little by the ventral one, the margin of the former being bent somewhat into the tube enclosed by both; behind, a little before the terminal part, they separate, and leave between them a slit broadening distally.

The number of terminal pieces is four.

Td and Tv are long, almost equally developed; their distal ends are not calcified, and do not reach quite to the end of the style. To the inner dorsal edge of Td is attached a piece Td_2, which

___

[1]) If we should suppose a skeletal part to be developed in the ventral wall of this pocket, it would in all respects be corresponding to the «covering-piece» r found in *Rhina*.

is a rather thick lamella only partly calcified. Also a piece Tr_2 is indicated as a pretty long, thin lamella, which does not calcify or only calcifies to a very small degree (see fig. 10 in the text); it is connected with the proximal end of Tr, stretches forward inside the «thorn», and is with the anterior end firmly united with the aponeurosis, on which the «outer lip-muscle», bespoken afterwards, acts; by the pulling of this muscle at Tr_2 and Tr the dilated terminal part is brought back, and the thorn T_3 situated between the two said pieces is laid. T_3 has more particularly the form of a claw, whose proximal part is head-shaped and rather soft, wrapped in the soft tissue connecting it with the adjoining pieces.

The muscular system. The $M.\ adductor$ is distally not sharply separated from the $M.\ dilatator$, as part of the fibres of the former passes into the superficial medial layer of the latter. The former muscle is quite woven together with the $M.\ extensor$, so that it is only by preparing from the dorsal side far into the large, proximal muscular mass that a considerable portion of fibres is found, originating from the basale, and having a direction common in the $M.\ extensor$.

The bag-shaped part of the $M.\ compressor$ is very long and rather thick; in the specimen examined by me, it is about 40^{mm} long, of which 23^{cm} are situated under the ventral skin before the pelvis[1]. The «outer lip-muscle» shows the peculiarity that in spite of the long ventral marginal cartilage it is prolonged covering the dorsal surface of the said cartilage until the terminal part, where it acts on Tr by means of the above mentioned lamellar indication of a Tr_2 in a similar manner, as this muscle acts in Sharks with a short ventral marginal cartilage.

The $M.\ dilatator$ only covers a very small part of the dorsal side of the appendix-shaft, by far the greater part of the dorsal marginal cartilage being covered only by the skin.

Selachus maximus (Gunnerus).

The appendix has been briefly mentioned by Sir Everard Home [2], somewhat more detailed by Blainville [3], but not originally by Pavesi [4], whose specimen, however, was a male; only in his second paper [5] does Pavesi briefly describe and draw (p. 353) the (undeveloped) appendix, and collects the whole literature treating of these organs, giving also in a table (l. c. p. 406) the dimensions that may be put together according to the obtained facts. The image of the appendix that is to be got from the literature, is upon the whole only imperfect. I have not found any particular mentioning

---

[1] The glandular bag contained only a little mucus, while the tube of the appendix, and the above mentioned pocket as well as the inside of the terminal part were all filled with an extremely viscid, milk-white mucus, which made the fingers exceedingly slippery and was difficult to get washed off; it contained numerous cells of different size and shape, with oval or round nuclei staining very readily.

[2] 1) An anatomical Account of the Squalus maximus etc. Phil. Trans., 1809, S. 207. 2) Additions to an Account etc. Phil. Tr., 1813. S. 230. Among other things the glandular bag is here mentioned as «a cavity between the skin and muscles of the abdomen, eleven feet long and two wide. The inner surface of this cavity is smooth, almost polished, and of a beautiful white colour; it contained a white mucus, extremely viscid and tenacious.»

[3] Mémoire sur le Squale Pélerin. Ann. du Muséum d'Hist. Nat., T. 18, 1811.

[4] Contribuzione alla storia naturale del Genere Selache. Ann. del Mus. Civico di Genova, vol. 6, 1874.

[5] Seconda Contribuzione alla Morfologia e Sistematica dei Selachi. Ann. del Mus. Civico, vol. 12, 1878. Besides the drawings quoted here (one after drawing is found only however a sketched outline, of evidently undeveloped appendices of a «Squalus maximus», in Carus und Otto: Erläuterungstafeln zur vergl. Anatomie. Part 5, pl. V, fig. VIII, 1840.

of the skeleton, but the following will, notwithstanding it defectiveness, show, that the structure of the skeleton is like that of *Lamna*.

In the museum of Copenhagen is found a pair of dried skeletons of these organs that have been got from a stuffed specimen (from California) of the length of 9^m 15^{cm} (27^1, Danish feet).

By long soaking the dried softer cartilaginous parts swelled so much that upon the whole they might be thought to approach to the shape of the fresh skeleton. Of the parts of the fin skeleton proper is only found a little, somewhat triangular piece, situated proximally at the dorsal end of the appendix-stem; it must be the piece β, which is accordingly (as presumably also the piece b_1) quite small as in *Lamna* (and the Scylliidæ)[1].

The appendix-skeleton has a length of about 1^{m}[2]; it may be doubted if all the terminal pieces have been preserved, but the principal features may be seen distinctly enough. The appendix-stem is calcified (the proximal end, however, not very much); the soft style is very long and rather broad, and reaches to the outermost end of the terminal part (comp. *Lamna*). The marginal cartilages are developed almost as in *Lamna*, that is to say, they join dorsally without forming a firm suture, the edge of the ventral one overlapping that of the dorsal one[3]. The number of terminal pieces (in the specimen in hand) is 3.

Td is short, not reaching to the end of the style; it is calcified for the greater part of its length, and has on the dorsal surface a furrow or groove f, wide before, where it passes into the large appendix-slit, while behind it becomes a narrow slit following the piece to the end[4].

Tv is only calcified anteriorly, otherwise it is a soft cartilage following the style just to the end. T_3

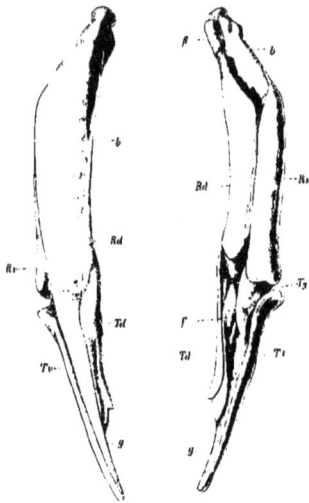

Fig. 11.   Fig. 12.

Fig. 11. *Selachus maximus*. The skeleton of the right appendage seen from the ventral side; much reduced.

Fig. 12. The same from the dorsal side. f furrow in the dorsal terminal piece. Both figures have been drawn after a dried skeleton. The position of T_3 is scarcely quite correct, and Tv is separated from its connection with the marginal cartilage Rv.

[1] Pavesi (1878, p. 378, fig. 12) draws the ventral skeleton of a young male; here is only seen the basale, and a very little developed stem-part of the appendix. As, however, in other Sharks the pieces, which I have here called b_1 etc. and β, are distinctly present in young ones, even in embryos, it is to be supposed that they have been overlooked here; in *Scyllium* and *Lamna* they are so small, that they are easily overlooked, if the skeletal parts are not cleaned of the soft parts with especial care.

[2] In the specimen of Blainville it was 3 feet long (the free part); the length of the animal was 29 ft. 4 inch.

[3] The words of Blainville l. c. p. 125 are: ils offroient en outre une fente ou sillon étendu dans toute leur longueur, mais dont la moitié antérieure, d'à peu près 14 ponces, étoit étroitement fermé par le rebord de deux cartilages très-serrés et qu'on ne pouvoit écarter qu'avec une très grande difficulté.

[4] It may possibly be this(?) slit, which is mentioned by Blainville l. c. p. 126 as a «sillon . . . . . beaucoup plus petit et plus étroit . . . . .» etc.

is claw-shaped and of a considerable size (in the specimen before me 16^{cm} long, and the broadest part 5,6^{cm} broad); with the exception of the proximal part it is completely calcified; according to the statement of several authors[1]) the point of it (in the developed organ) projects through the skin.

Besides the three terminal pieces seen in my figures, I think it probable that one more has been found, a Td_4 as in *Lamna*. I found this opinion in the first place on the words of Blainville (l. c. p. 126) that besides the claw there is un autre cartilage, un peu aplati, occupant le milieu du tiers antérieur de cette gouttière (i. e. the furrow of the terminal part); celui-ci étoit mobile presqu'en tous sens, mais entièrement renfermé dans un repli de la membrane interne qui se prolongeoit, libre et flottante, jusqu'à l'extrémité posterieure du sillon. Next I found the above stated opinion on the description (1878, p. 352) and drawing in woodcut (fig. 3) of the (undeveloped) appendix given by Pavesi: nella metà apicale offrono un pezzo mediano lanceolato, rialzato e piano, con fenditure laterali. Questo superficie non ha traccia di sperone corneo. Later (p. 405) it is said of this piece that it is only a thickened dermal fold, not to be confounded with the spur [2]. The dermal fold mentioned by these authors, no doubt corresponds with that one which in *Lamna* contains the piece Td_4. But what is the fenditure laterale of Pavesi? According to the figure it must be situated on the medial side of the organ, that is to say, it is presumably the sillon.... beaucoup plus petit et plus étroit of Blainville; and thus it must be supposed to be the one seen in the skeleton, fig. 12 *f*, and not a *pocket* like the one described above in *Lamna*, because this latter is situated before the terminal part, and accordingly would be seen on the part called by Pavesi la meta basale.

Rhinidæ.

Rhina squatina (L.).
(Pl. II, fig. 24—27.)

In a specimen of the length of 1^m and a breadth across the pectorals of 0,59^m, the part of the appendix free of the fin is 8^1,^{cm} in length; from the foremost beginning of the slit the length is

[1]) Shaw: General Zoology V, pt. II, Pisces, 1804, tab. 149 (in the text nothing is found about it); the figure is certainly bad, and the appendices can scarcely ever have that appearance, but are, to use the words of Pavesi (1878, p. 404). trasformate in sorta di gambe dall'imaginoso disegnatore. Blainville gives it to be 7 inches long, but covered by soft tissues except 1/2 inch, which m'a paru comme cornée et libre au bord supérieur et extérieur de l'appendice. Home speaks of it as a strong, flat, sharp, bony process, five inches long, which moves on a joint, and the bone projects an inch and a half beyond the skin, like a spur (1809, p. 207); in the later addition is only said; the spur bears a striking resemblance to that of the male ornithorynchus paradoxus. Lesueur: Description of a Squalus etc.; Journ. Acad. Nat. Hist. Philad. II, part II. 1822, p. 349; Mitchill in Dekay: Natural History of New York, Zoology, part IV, Fishes, 1842, p. 358: From and between the anal fins, two legs project five feet in length, and are terminated by a claw tipped with horn. Van Beneden: Un mot sur le Selache (Hannovera) aurata du crag d'Anvers; Bull. Acad. Roy. de Belgique, 2 Série, vol. 42, 1876, draws a sketch of the appendices with the spur from a stuffed specimen in British Museum, and shows that these spurs are (as well as the gill-rakers) found as fossils in tertiary strata. Before I knew this fact and the paper by van Beneden, I have expressed, in a lecture given in the Society for Natural History in Copenhagen (March 1897), the conjecture that the very hard, dentine-like terminal pieces of the appendices of Selachii might exist as fossils, and indicated that perhaps some of the *ichtyodorulites* were not dermal teeth (spines) but such skeletal parts; by turning over the work by Agassiz on fossil fishes I have, however, not been able to find any drawing, to which this conjecture might be applied.

[2]) Pavesi himself thinks the presence or absence of this latter to be dependent on the age of the animal, and not to indicate a difference of species, and it is now beyond all doubt that this opinion is quite correct. All other species of Sharks that are provided with a similar spur (as *Acanthias, Spinax, Somniosus* a. o.) show that this piece is formed hidden in the skin, and is not uncovered until it has reached a considerable degree of development, contemporary with the organ as a whole having altered its shape and dimensions.

11,5ᶜᵐ (from the foremost edge of the pelvis to the hindmost point of the appendix the length is 24ᶜᵐ) [1]); the largest breadth is found somewhat above the hindmost point of the fin-membrane, and is about 3,5ᶜᵐ. The ventral side is flat, covered with dermal teeth till the naked terminal part (4,3ᶜᵐ long, 2,1ᶜᵐ at the broadest spot, at the base); these dermal teeth are flat, and form a complete mosaic, only a little rough to the feeling, when rubbed with the finger towards the base, and it is of quite the same nature as that covering the ventral side of the rest of the fin [2]); the whole surrounding of the cloaca as well as the space from there to the foremost lateral corner of the fin is naked (with a few scattered groups of teeth); further-more a naked stripe stretches from the hindmost inner edge of the fin-membrane, where the fin is laid against the appendix, some way on the ventral side of the fin. Both edges as well as the whole dorsal side of the appendix are naked.

Fig. 13.

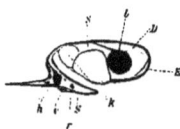

Fig. 14.

Fig. 13. *Rhina squatina.* The left appendage from the dorsal side, reduced. The part from *a* to *b* has been cut up in prolongation of the appendix-slit to open this latter so much as to get a view of the gland *k.* *l'* is part of the entrance to the pocket between *Tb* and *v.*

Fig. 14. A transverse section through the same appendage after the line, marked with • • in fig. 13. *b* the transverse section of the appendix-stem. *D Musculus dilatator.* *E Musculus extensor.* *S Musculus compressor.* *k* the gland. *v* blood-vessels. *b* horny filaments. *r* the end of the last ray.

The somewhat triangularly pointed terminal part shows, when seen from the dorsal side, the appendix-slit (*af*) situated near the lateral edge; a rather large, foliace-ous fold of the skin, which includes the skeletal piece *Td₂*, originates from the dorsal lip of the slit; the proximal end of this fold stretches into the (half-) tube formed by the distal ends of the marginal cartilages. Accordingly there will, on both sides of this plate, be an outlet from the canal of the glandular bag in the broad part of the appendix; the real continuation, however, of this canal is here, as everywhere, situated in the terminal part laterally of the said fold (at *af*). If the fold is thrown back, there will in the lateral margin of the appendix-slit proper be seen a rather large

[1]) Müller & Henle l. c. p. 99 state in characterising the genus: ›Die Anhänge des Männchen klein und weich‹, which, as we have seen, does not apply to adult animals, and consequently is of no value as a characteristic of the genus.

[2]) This dental mosaic on the ventral side is quite different from that of the dorsal side, where the dermal teeth project as small thorns from the thicker skin. The words of Müller & Henle in characterising the genus: Schuppen konisch in eine Spitze endigend, zerstreut‹ are then only to be applied to the dorsal side. I have not upon the whole found any mention made of the dermal teeth of the ventral side, only expressions to the effect that the abdomen is more or less smooth.

aperture (at *l'*) leading into a pocket between the ventral terminal piece (*Tv*) and the ventral covering piece *v*.

If the lateral lip of the appendix-slit above the terminal part is lifted, a thick g l a n d u l a r b o d y is seen protruding from the medial side of the canal, in which feature this species differs from all other Sharks I have had the occasion to examine. This gland will be more particularly mentioned afterwards.

T h e s k e l e t o n. Between the basale and the appendix-stem three short pieces (b_1, b_2, and b_3) are found, each bearing one of the three hindmost rays (the two hindmost of these rays are terminally quite coalesced for a long way). At a first glance the piece *β* seems to be wanting, but a closer examination shows it to be present, represented by a little cartilage, arising from the lateral hind corner of b_3, and joined by a particular articulation to the proximal end of the appendix-stem *b*. Contrary to what commonly is the case in the Sharks, the piece *β* does not here articulate proximally with the basale.

The a p p e n d i x - s t e m (*b*) is long, considerably longer than the basale (the ratio is ³⁄₂), round (with the exception of the proximal part where it dorso-ventrally is somewhat flattened); the style is long, not, however, reaching to the hindmost point of the terminal part. The marginal cartilages are s h o r t, and are only found at the distal part of *b*; contrary to what commonly is the case in the Sharks, the v e n t r a l marginal cartilage is the one reaching most forward. The general shape of these cartilages, I think, may be seen with sufficient clearness from the figures. The ventral marginal cartilage bends towards the dorsal one with a plate similar to that found in many other Sharks, but does not quite reach it. But this plate is here in a peculiar way hollow, being behind split into two lamellæ receiving between themselves the proximal end of the piece *Tv*; this piece, then, projects into the ventral cartilage, quite covered, until the point marked * in fig. 25 ¹). The hindmost end of the inner one of these two lamellæ protruding very much, the appendix-slit is by its transition to the terminal part straightened to an extraordinarily narrow passage.

The n u m b e r of t e r m i n a l p i e c e s must in reality be taken to be four; but to these four is added a good-sized, ventrally situated piece, *v*, rounded in a scutiform manner, and partly covering the terminal part (see fig. 24) behind the ventral marginal cartilage. This piece has developed in the aponeurosis which, in the Sharks hitherto mentioned, encloses the terminal part, and it serves like this aponeurosis for inserting the large *Musc. dilatator*. If this piece *v* is removed, the ordinary terminal pieces are easily recognized: *Td* which is rather broad, flat, with a thickened edge medially (which edge follows the style closely, but reaches a little further backward), and a sharp and thin edge laterally; Td_2 proximally joined to the foregoing piece, is a broad, but thin, and but slightly calcified lamella. *Tv* is of a very peculiar shape, thick and solid, ventrally rounded, dorsally, towards the slit, deeply hollowed in a spoon-like shape; its proximal end, as already mentioned, passes its articulation with *Kv*, and enters between the two lamellæ of the overlapping plate; with the proximal end articulates, completely hidden, a little calcified piece representing the thorn or spur , T_3; this latter piece (see fig. 27) is proximally irregularly head-shaped, and from this thick part a thinner one

1) If in *Acanthias* or *Spinax* the projection of the ventral marginal cartilage, mentioned at p. 28, was more developed, and proximally prolonged, a similar state of matters might be the result.

projects, bent at its rise, but otherwise straight, cylindrical, and rounded posteriorly. It is necessary in order to get a view of this piece T_3, and to isolate it together with Tr, to cut away part of the outer lamella of the ventral marginal cartilage.

The muscular system. The *M. adductor* shows no deviations from the common type; the *M. extensor*, on the contrary, shows the peculiarity of being divided into two independent muscles (comp. *Torpedo*), an inner (medial) one, and an outer (lateral) one, bordering on each other, and both originating from the basale; the inner or foremost one arises rather far forward on the basale alone, runs, like the *M. extensor* in the Greenland Shark, across the appendix-knee, covering as a thin plate part of the dorsal side of the *M. dilatator*, and ends quite posteriorly, at the terminal part. The outer or hindmost extensor arises behind the foregoing one, not from the basale only, but also from the pieces b_1, b_2, b_3, and it is attached to the appendix-stem immediately behind the knee .

The *M. dilatator*, as is commonly the case, encompasses the appendix-stem from the dorsal marginal cartilage to the ventral one; the lateral part of it arises forward on the ventral side of the basale and the short pieces following this latter (comp. the Rays); behind its chief portion is attached to the ventral covering piece r.

Of the *M. compressor* the bagshaped part is rather short, and does not nearly reach to the pelvis, but otherwise it agrees with the one found in other Sharks. The outer lip-muscle is very powerful as in the other Sharks with a short Rr, and is attached posteriorly chiefly in the aponeurotic covering of the piece Tr.

The secreting part of the glandular bag shows in its foremost part the same relations as in the other Sharks; but in the part which is situated in the shaft itself, a large glandular body (see fig. 13 and 14) has been developed on the ventral side. The presence of this gland may already be guessed by the peculiar exterior of the appendix-shaft; its proximal part shows, when seen from the ventral side, a peculiar convexity, by which the organ gets a contour not unlike that of a human leg with a large calf. The glandular body reaches before quite to the beginning of the bagshaped part, that is to say, much farther than the appendix-slit itself, so that it is necessary to cut up some way (see fig. 13) in order to get a view of the foremost end; it is a little tapering behind, and reaches to the terminal part. A slight, longitudinal furrow is found on the free (dorsal) surface about the middle, and on the edges of this furrow are situated two series of large glandular outlets; a great number of similar outlets are also found laterally of the furrow, in pretty irregular groups; to the medial side of the furrow are also some such openings, but apparently in much smaller number. When the gland is pressed an abundance of mucus will appear as stoppers in the said outlets. The glandular body is composed of dichotomously branched tubes, quite similar to those found in the Rays, and with quite similar large secreting cells; but they are here grouped in a somewhat different manner as a consequence of the outlets of the gathering ducts being spread on a much greater space. The glandular body in *Rhina* furthermore deviates from that of the Rays by its ventral position in the shaft [1]), and by not having the special muscular coat developed as in those; the part of the *M. compressor* situated at the gland will very likely be able to act in a similar way, possibly only with less force, in *Rhina*.

---

[1]) In *Torpedo, Narcine, Rhinobatus* and *Trygon* the dorsal glandular body of the bag is continued throughout the shaft with the same structure as in the bag, but reduced in bulk, and situated along the ventral marginal cartilage. If in one of these Ray-forms the part of the gland situated in the bag be supposed absent, and the part in the shaft displaced

The peculiar mixture of Shark-like and Ray-like characters that, as it is well known, is found in *Rhina*, is accordingly increased by several features in the appendages of the male, which features by the ventral covering piece and the pocket, situated below it, with entrance from a side-slit, and partly also by the glandular bag, recall those in the Rays (*Torpedo*, *Narcine*, *Rhinobatus* and *Trygon*), while most of the other features are those common in other Sharks.

Cestraciontidæ.

Heterodontus (Cestracion) Phillipi (Cuv.).

The skeleton has been described by Gegenbaur[1]. Between the basale and the appendix are found two pieces (b_1, $b_2 = \beta$, β', l. c. fig. 18, 19) that bear no rays; the piece β is well developed (l. c. *b* fig. 19). The chief piece of the appendix is provided with two (rather long?) marginal cartilages (the boundary lines of which cannot be seen in the figures of Gegenbaur, as he has not understood the marginal cartilages to be particular pieces), of which the ventral one has a dorsally bent plate (l. c. fig. 19, *a*); the stem is prolonged into a long style reaching almost to the end of the terminal part (l. c. fig. 19, 20, *i*). The number of terminal pieces is four: *Td* (= l. c. fig. 19, 20, *o*), Td_2 (= l. c. *u*), which, as is often the case, is proximally prolonged into the appendix-slit; *Tr* (= l. c. *r*), as commonly, stronger and thicker than the others, and finally T_3 forming a short thorn. Gegenbaur has correctly seen the homologies of these pieces with those in *Acanthias*, where, however, he has not seen the piece Td_2 (= *u* in *Heterodontus*). Of these terminal pieces the piece T_3 is said (l. c. S. 452) to be hard, while the others, though fully developed, are still cartilaginous.

Notidanidæ. [2]

Chlamydoselachus anguineus Garman.

Günther[3] has briefly described the appendages and their skeleton, and given figures of them. Only a third part of the length of the appendages is free of the fin, „as is the case in the *Notidanidæ* generally", and there is no notch in the hindmost fin-edge, between the membrane and the appendage. Between the basale and the appendix-stem there are „three rudimentary and one larger intermediate cartilages" (b_1, b_2, b_3, b_4?), none of which bears any ray. To judge by the figure, there is no piece β;

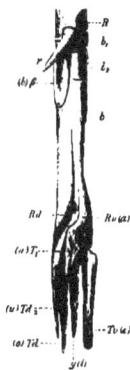

Fig. 15. *Heterodontus Phillipi.* The skeleton of the right appendage. After Gegenbaur(l.c. fig. 19), somewhat reduced. The letters placed in parentheses are those used by Gegenbaur. *r* the last ray.

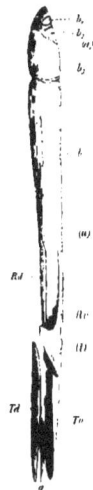

Fig. 16. *Chlamydoselachus anguineus.* The skeleton of the right appendage. After Günther, somewhat reduced. The letters in the parentheses are the original ones.

along the ventral marginal cartilage to the lateral surface of the appendix-stem, we should have a similar state of matters as in *Rhina*. There can scarcely be any doubt that the gland in this Shark and in the Rays — in spite of the difference of position — are in reality homologous. Furthermore the glandular bag in younger stages of the Rays seems to pass through a stage of development, in which there is, also as to the exterior, a conspicuous similarity with that of the *Rhina*, without any conspicuous longitudinal furrow etc. (see later under *Raja batis*).

[1] Über die Modificationen etc. 1870, p. 450, Taf. XVI, fig. 18—20.

[2] I regret very much that my efforts to get ventral fins with developed appendages of *Hexanchus* or *Heptanchus* have been in vain. The figure of the skeleton of *Heptanchus cinereus* Ag. given by Fritsch in Fauna der Gaskohle etc. Böhmens, vol. 3, 1895, p. 43 is quite useless. From this figure appears only that at least the two terminal pieces *Td* and *Tr* are found; what Fritsch calls the ·sporn· is the last ray (or rather the two last, coalesced ones). The *r* of the figure, I suppose to be the piece β, and it is certainly not the "Letztes Glied des Hauptstrahles". The figures of the structure of the appendages in the extinct *Xenacanthidæ*, given by Fritsch as well in his chief work as in several articles in the Zool. Anzeiger (non l. gr. I think to be justified in designating as unreliable) but by means of the published figures alone the real structure cannot be determined.

[3] Voyage of H. M. S. Challenger, Zool., Vol. XXII, 1887, S. 2, Tab. LXIV, fig. *C*, *D*, *E*.

as, however, no figure is given of the part in question, seen from other sides, I cannot regard this absence as quite certain. The long appendix-stem is prolonged to a style reaching to the end of the terminal part. The marginal cartilages (not understood by Günther to be particular pieces) appear to have mainly the same structure as in the Spinacidæ; the ventral one has the usual overlapping plate (l. c. T. LXIV, fig. D, D', I). The number of the terminal pieces is only two (if not a piece T_3 has been overlooked or removed by the preparation?) viz. Td and Tv, both hard and calcified, Tv being as usual largest and broadest.

Carchariidæ.

Mustelus antarcticus Gthr.

A pair of dried skeletons of ventral fins with appendages in the Zoological Museum at Copenhagen.

Between the basale and the appendix-stem one rather small piece b_1 bearing the last ray, which is partly coalesced with the last but one; a distinct, well developed β that seems to have been

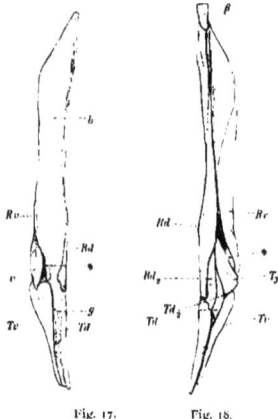

Fig. 17.   Fig. 18.

Fig. 17. *Mustelus antarcticus*. Skeleton of the right appendage, seen from the ventral side; about the natural size. *v* the ventral covering piece; * a spoon-like hollow in the ventral marginal cartilage.

Fig. 18. The same preparation from the dorsal side; * the bottom of the hollow in the ventral marginal cartilage, protruding into the appendix-slit. Both figures have been drawn after the dried skeleton.

triangular. The appendix-stem is prolonged to a long, soft style reaching almost to the hindmost end of the terminal part. The marginal cartilages stretch over the hindmost two third parts of the chief piece, the dorsal one reaching farther forward than the ventral one; the edges of their folded parts join dorsally, leaving between them only a narrow slit. To the distal end of the dorsal marginal cartilage is added a foliaceous, slightly calcified piece, homologous with the piece Rd_2 mentioned in *Pristiurus*.

The number of terminal pieces is 4 (÷ 1).

The two of these pieces that as usual follow the style, and together with it form the walls of the hindmost part of the appendix-slit, Td and Tv, are well calcified, lengthened, pointed, and Td a little longer than Tv. To the lateral edge of Td is proximally added a foliaceous, slightly calcified piece Td_1, forward stretching under the piece Rd_2 into the appendix-slit as in several other Sharks. Finally there is a rather large, flat, triangular, posteriorly taplike piece T_3 that, however, does not not appear to have projected through the skin as a spur.

Besides these real, typical, terminal pieces still a special piece, v, is found, which I take to be corresponding to the one marked with the same letter in *Rhina*, and accordingly to have arisen from the aponeurosis of the *Musc. dilatator*; here in *Mustelus antarcticus* it covers a peculiar, rather deep, spoonlike hollow on the ventral side of the end of the marginal cartilage K_2 (on the dorsal side the corresponding spot is seen protruding into the appendix-slit, at * in fig. 18).

The descriptions of different Sharks given in the preceding section will have shown that the common type in the skeleton is clearly conspicuous; the single secondary skeletal pieces may vary pretty much, as to their form, but their homologies are easily and surely demonstrated. Although the mentioned species of Sharks cannot be said completely to represent the Sharks upon the whole, yet they belong to so many different families that we may be justified in coming to the conclusion that the skeletal structure of the appendix is in the Sharks rather simple and easily explained. This, however, can in no way be said of the Rays in general; here, especially in the genus *Raja*, may be found particularly complicated structures varying to a high degree even from species to species; and as the *Raja*-species are those that have been especially examined by earlier authors, it will easily be understood that so few general results have hitherto been obtained. If, however, by means of the Sharks we have got a clear understanding of the characteristic common features, it will not be so very difficult to point out these features also in the *Raja*. It is, however, an obvious supposition that other forms of Rays than *Raja* will approach more nearly to the Sharks, and such forms will most likely have to be sought among the shark-like Rays, as *Pristis*, *Rhinobatus*, *Torpedo*, etc. Through the kindness of Professor Lütken I have from our museum obtained the material of the two last-named genera, and of *Torpedo* I have also got some pairs of ventrals from Napoli.

Batoidei.

Torpedinidæ.

Torpedo marmorata Risso.
(Pl. III, fig. 28—31.)

The appendix, like the whole fish, is naked, flattened, with tolerably parallel sides, the terminal part oval, distinctly marked off from the shaft by a slight constriction. The appendix-slit runs on the dorsal side quite straight, nearest to the lateral edge until the hindmost half of the terminal part where it suddenly bends to the medial side, and with a curve reaches to the end. On either edge of the terminal part is seen a lengthened slit, posteriorly widening somewhat like a buttonhole; either slit leads into a blind, pocket-like bag, inside the later mentioned covering piece v^1). None of the enclosed skeletal pieces are naked.

In a specimen of the length of 29^{cm}, a breadth of 17^{cm} the following measures are found for the appendix:

¹) Petri l. c. pl. XVI, fig. 4 gives a — neither good nor exact — figure of the appendix, seen from the dorsal side.

Length from the foremost end of the slit to the extremity . . 40^{mm}

   —   of the free part . . . . 17^{mm}

The greatest breadth . . . . ca. 10^{mm}

The length of the terminal part . . . . 15^{mm}

In another specimen of the same length and a breadth of 19^{cm} the appendages were a little shorter and broader, but otherwise as fully developed.

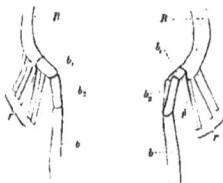

Fig. 19.   Fig. 20.

Fig. 19. *Torpedo marmorata*. Part of the skeleton of the right ventral, from the ventral side. *r* the hindmost rays partly cut off. Natural size.
Fig. 20. The same preparation from the dorsal side.

The skeleton. Between the basale and the appendix-stem are found two pieces b_1, b_2, of which the former is the shorter one (seen dorsally it is much shorter), and bears the two hindmost rays (the last but one is partly borne by the basale also). The piece β is long, longer than b_2, flat; proximally it articulates with b_1, runs along b_2 without touching it, and articulates distally by an oblique articulation with the appendix-stem [1].

The appendix-stem is rather straight, calcified in the surface until the terminal part, where the mosaic of the surface suddenly ceases, and the outermost part of the stem is soft, which soft part, chiefly of the same breadth as the harder one, thus corresponds to the style, and reaches to the end of the terminal part, ending with a broadly rounded, convex edge.

The marginal cartilages are both calcified, but very unequally developed, by which the whole appendix gets a peculiar asymmetric appearance; the ventral one [2] reaches a trifle longer forward than the dorsal one, but backward it ceases far before this latter; the dorsal cartilage is by a longitudinal furrow apparently divided into two pieces of which the lateral one begins forward about the middle of the ventral cartilage, and stretches backward about as far past the hinder end of this latter (see fig. 30, 31); this part of the dorsal marginal cartilage is on the ventral side hollowed in a trough- or groove-like manner; the medial part of this cartilage, especially the foremost part of it, is slightly calcified, membranous, and is placed like a cover over the appendix-slit, so that a narrow slit is left between its outer edge and the ventral marginal cartilage.

---

[1] Petri has quite misunderstood the relations of these skeletal pieces, and has upon the whole been very unlucky in his explanation of the skeletal pieces in *Torpedo*. Already his beginning: Das Skelet von Torpedo besitzt nur sehr geringe Ähnlichkeit mit dem der vorher beschriebenen Arten , promises nothing good; and his description of the terminal parts shows that he has not understood them at all. He has correctly seen that between the basale and the appendix are situated two pieces: fig. 4 D, *b'*, *b''*, but their length and position is given less correctly in the figure, which is, like all his figures, rather bad; but then he has completely overlooked the piece β as a separate skeletal part taking it to be a process on the appendix-stem: Nach vorne entsendet dasselbe (i. e. the appendix-stem) an der medialen (i. e. the dorsal) Seite neben dem zweiten und dritten Glied des Basale entlang einen Processus, welcher mit dem ersten Basale am hinteren Ende noch in Verbindung steht . As a homologon to the piece β in *Acanthias* (Petri's fig. 5 *r'*) he takes a little piece (fig. 4 D, *r'*), which is said to bear the two last rays and to have originated from a coalescing of the proximal joints of those: es ist dieselbe Concrescenz, wie ich sie bei *Acanthias* beschrieben habe . In *Acanthias*, however, the piece β *u'* in Petri bears no rays, as, after all, it never does. The little piece which Petri has seen in *Torpedo* (and in the figure marked *r'*), is only an uncalcified corner of the basale itself, projecting over the two last rays, but it does not bear the rays.

[2] The marginal cartilages are partly correctly seen and determined by Petri as : Rinnenknorpel : Fig. 4 D and E, *e* and *l*; F. *l*; (and pl. XVII, fig. 4 B, *l*); the skeletal piece, however, interpreted by Petri as a dorsal marginal cartilage and marked *l*, is only the firm lateral part of the dorsal marginal cartilage; the thin, cover-shaped part seems to be removed, except on fig. 4 B, representing the muscles.

The number of the terminal pieces is three; but to be able to see these pieces, and correctly to understand their relations to the marginal cartilages, it is necessary to remove a large ventral covering piece, homologous with that mentioned in *Rhina*, and marked *v*, because the whole ventral side of the terminal part is hidden as by the half of a thimble (the contour of which may often be distinguished through the skin). The foremost edge of this covering piece is incised in an irregularly heart-shaped manner, the one — lateral — corner being produced much farther than the other; it is strongly rounded from side to side, and the surface is dotted with small holes. The edges of this piece have on each side a peculiar bend, and immediately behind this the covering piece is firmly joined to the specially thickened hinder edges of the two terminal pieces *Td* and *Tv*; by the said bends and the ventral concavities of the covered terminal pieces, the peculiar side-slits and the walls of the before mentioned / pockets are formed. The *M. dilatator* is attached to the fore edge of the covering piece, and by the firm connection between the covering piece and the outermost point of the terminal pieces the action of the muscle is transmitted to that point.

The dorsal terminal piece, *Td*, is quite short, somewhat convex towards the appendix-slit; the ventral side on the contrary is deeply concave; together with the corresponding marginal cartilage which, as we have seen, is also hollow, it forms a complete trough, in which the medial pocket is situated. The ventral terminal piece, *Tv*, has the double length, is likewise rounded on the dorsal side, hollow on the ventral one, and forms with the covering piece the lateral pocket. Finally a piece *T₃* is added to the hindmost end of the ventral marginal cartilage, and to the proximal part of *Tv*; it is shaped as a slightly bent, round thorn, almost hidden inside of the lateral edge of the covering piece [1]).

The muscular system. The *M. extensor* is divided into two parts reminding of the state in *Rhina*. The foremost one originates from the foremost half of the basale, runs over the knee, and is attached immediately below this to the *M. dilatator*, and partly to the distal end of *β* and of the appendix-stem; the hindmost one originates from the other, hindmost, half of the basale and from the following pieces (also from *β*), passes with its distal part under (i. e. ventrally of) the foregoing one, and is continued rather directly in the medial part of the *M. dilatator*.

M. compressor. The bag-shaped part of this muscle, situated on the ventral side, is very small; properly speaking it is confined to a ventrally rounded swelling between the two hindmost rays and the parts of the stem-skeleton, b_1 and b_2; accordingly there is no part projecting across the ray muscles, the foremost contour of the bag being bordered by the skeleton (the pieces b_1 and b_2), the outermost one by the hindmost ray muscles. The ventral fibres of this little bag run obliquely backward towards the hindmost ray, almost in continuation of the hindmost fibres of the *M. adductor*. The part forming the outer lip-muscle, is rather powerful and long, originates forward from the piece *β* (foremost

[1]) Petri, as already mentioned, is quite wrong as to the terminal part; he takes the dorsal terminal piece *Td* to be the terminal joint of the appendix-stem (l. c. p. 308, fig. 4 F, *b'''*); the ventral terminal piece and the thorn *T₃* become »zwei kleine verkalkte Spangen«, borne by the hindmost part of the ventral marginal cartilage to which they are added »radienartig« (fig. 4 E and D, *sp''* [= *Tv*], *sp'* [= *T₃*]). Only the piece *v* (fig. 4 E, F, D *sch*) has been tolerably correctly described and interpreted by Petri. Thus he has had no idea of the homologies of the terminal pieces with those in *Acanthias* and *Scyllium*, not to speak of *Raja*.

also from b_1), and the fibres pass obliquely backward towards the last ray and the ventral marginal cartilage. On the inside of this latter the muscle continues somewhat farther back, tapering, and with the fibres running straight backward. The whole muscular wall may pretty easily be separated from the two hindmost rays to which its fibres are not really attached; when thus separated from the rays the whole muscular bag shows a rather strong resemblance to *M. compressor* in *Chimæra*.

The *M. dilatator* is with the whole of its hindmost end attached to the covering piece *v*, and thus, by means of the firm connections of this piece with the distal end of the terminal part, it acts on the movable portions of the terminal part[1].

The glandular body is pigmented, and its longitudinal furrow runs a little obliquely; it has before been briefly described by Leydig[2]) and by Petri (l. c. p. 22), but none of these authors mention the peculiar fact, in comparison with the *Raja*-species (and accordingly it is to be supposed that it has not before been seen), that the glandular body, continually tapering posteriorly, stretches throughout the shaft quite down to the terminal part. In the shaft it follows the ventral marginal cartilage, and is here for some way along this skeletal piece surrounded by the outer lip-muscle, the latter appearing as the continuation of the dorsal muscular wall of the *c*bag. The longitudinal furrow, and the openings in it follow the glandular body quite to its hindmost end.

Torpedo oculata Bélon.

In this species the features are principally as in *T. marmorata*.

In a specimen of a length of 30^{cm}, a breadth of 19^{cm}, the following measures were found:

Length of the appendix . . . . 43^{mm}
- - - slit . . . . 36^{mm}
- - part free of the fin . . . . 17^{mm}
— - - terminal part . . . . abt. 14^{mm}
Breadth of the appendix . . . . 7—8^{mm}

Narcine sp.
(Pl. III. fig. 32. 34.)

A badly preserved specimen, the species of which is difficult to determine, measures in length 24,5^{cm}, in breadth 12^{cm}:

The appendix from the beginning of the slit to the hindmost point is . . . . 27^{mm}
- free part. . . . . 14^{mm}
- length of the terminal part. . . . . 10^{mm}
- breadth of the -- . . . . 5^{mm}

---

[1] Petri has given a very imperfect description of the muscles of which he only mentions *M. dilatator*. (Comp. l. c. fig. 4 B and C.)

[2] Beiträge zur mikroskopischen Anatomie und Entwickelungsgeschichte der Rochen und Haie, 1852, p. 86.

The exterior of the appendix is chiefly as in *Torpedo*; we find here the same marginal slits (and pockets), but the medial one is with the buttonhole shaped, distal part situated quite on the dorsal side, and the lateral one is turned a little ventrally; next it is to be remarked that the appendix-slit in the whole of the free part of the organ is lying quite laterally, the dorsal lip overlapping it quite to the outer edge like a cover, much broader than in *Torpedo*; corresponding to this the portion of the terminal part containing the piece *Td* is folded quite over the hindmost part of the appendix-slit. The glandular bag and its inner gland are relatively more strongly developed than in *Torpedo*.

In the skeleton, notwithstanding the principal conformity with *Torpedo*, several peculiar features are found. Between the basale and the appendix-stem also here two pieces, b_1 and b_2, are found, the former short, the latter longer; b_1 bears the last ray at the connection with the basale; the last ray but one seems to me only to articulate with the basale. The piece β is totally wanting. The appendix-stem is long, considerably longer than the basale $+ b_1$ and b_2; it is calcified in the surface excepting the short end-style,

Fig. 21.

Narcine sp. The right appendage from the dorsal side. Natural size. *I* the opening of the medial pocket. *Rd*, *Td* skeletal parts covered by the skin.

which distally becomes broad, flat, with rounded hindmost contour. The marginal cartilages are thin, rather short, and occupy only the hindmost half of the stem; the ventral one shows the same features as in *Torpedo*; also here it reaches a little farther forward; the dorsal one, when seen from the dorsal side, forms a broad, ovate leaf tapering proximally, and continuing as an uncalcified band quite to the articulation of the appendix-stem with b_2; its lateral edge reaches to the free edge of the ventral marginal cartilage; seen from the ventral side it is hollowed in a trough-like manner as in *Torpedo*; but distally it does not nearly reach so far as in the latter, and consequently the two marginal cartilages do not end so obliquely of each other distally (this fact seems also to imply a greater mobility of the terminal part in *Narcine* than in *Torpedo*).

The number of terminal pieces is three, to which is to be counted a quite similar ventral covering piece *v* as that in *Torpedo*. As already mentioned, *Td* is folded to the dorsal side, and has apparently a shape deviating considerably from that found in *Torpedo*; a closer examination shows however that this deviation to some degree is due to the position; on the medial-dorsal side the piece is hollowed in a groove-like manner (for the medial marginal slit); otherwise it is flatly rounded on the outer surface, concave towards the appendix-slit, with a sharp lateral, convex edge.

Tv is short, oval, rounded towards the appendix-slit (as in *Torpedo*) concave on its outer surface, by which, together with the covering piece *v*, it forms the hollow for the lateral pocket. T_3 is somewhat s-shaped, tapering to both ends, little and slender. The covering piece is chiefly as in *Torpedo*; also here we find on both its margins curvatures destined (especially on the lateral margin) to form the button-hole shaped opening of the pocket together with the terminal pieces *Td* and *Tv*. The foremost lateral corner appears independent as a very small *v'*.

The muscular system is as in *Torpedo* with the exception that the part of the *M. compressor* that may be seen on the ventral side, is relatively larger, and laterally spreads somewhat more, and that the *M. extensor* is single. The glandular body is prolonged into the shaft quite as in *Torpedo*.

Rhinobatidæ.

Rhinobatus columnæ Bonap.

(Pl. III. fig. 35—37.)

In a specimen of the total length of 85,5^{cm} and a breadth of 26^{cm} across the pectorals the appendix had a length of 11,5^{cm}.

The other measures were:

from the fore end of the slit to the hindmost point of the appendix	7,4^{cm}
the part free of the fin	6,2^{cm}
Length of the terminal part	1,9^{cm}
Breadth - - -	0,8^{cm}

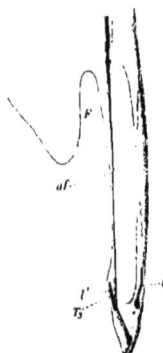

Fig. 22.

Rhinobatus columna.
The left appendage from the
dorsal side. Natural size. *F*
the fin-membrane. *l* the
opening of the lateral pocket,
l of the medial one.

The appendix reminds of that in *Torpedo*, being flat, and having (on the ventral side) the terminal part marked off by a slight constriction; but it is considerably longer, with a long free portion, and a relatively short terminal part, and so it has a rather slender, elegant appearance. The ventral side is all covered with the same dense mosaic of fine, flat dermal teeth as the other parts of the belly; only the very outmost point (1^{mm}) is naked and soft. On the dorsal side the region around the foremost part of the appendix-slit is naked as well as the whole part before the slit, which is normally in contact with the belly, this having also a corresponding naked spot; furthermore the naked part stretches as a small stripe backward close to the medial edge until the terminal part, the dorsal side of which is quite naked; otherwise the dorsal side is covered with teeth, and has the darker colour of the back of the animal. The appendix-slit is situated quite close to the lateral edge, only in the terminal part it bends towards the middle. On the dorsal side of the terminal part are seen two «marginal-slits» leading into pockets, quite corresponding to those in *Torpedo* and *Narcine*; the medial one is the longer; in the lateral one the outermost part of the piece T_3 is seen freely protruding from the skin as a shining thorn.

The skeleton. Between the basale and the appendix-slit are found four pieces: b_1, b_2, b_3, b_4; the last one being the longest, next to that follows b_1 which is also the broadest, but they are all more long than broad; b_3 and b_4 bear no rays, b_2 bears the last ray, b_1 the last but one (and, in connection with the basale, also the last but two). The piece β is exceedingly long; anteriorly it articulates with the dorsal side of b_1 close to the basale, then it stretches along the other pieces, b_2 b_3 b_4, and next articulates with the appendix-stem b in a long joint, reaching as far past the articulation of b_4, as the length of this piece b_4 itself; it reaches quite to the dorsal marginal cartilage. It is narrow and flat, distally a little bent. The appendix-skeleton is of a length about equal to the basale + b_1 . . . b_4, the stem is rather slender, and ends in a style of a similar shape and nature as in *Torpedo*, only it is here relatively very short. The marginal cartilages are very long; both of them begin almost at the same point forward, the ventral one, however, a little before the dorsal one, and they end at the style in an inverse ratio, that is to say, the dorsal one reaching a little farther backward. On the greater part of their external surface they are very closely connected with the teeth-covered skin, so that they only with difficulty can be separated from it. The distal part of the dorsal marginal cartilage is, quite as in *Torpedo*, ventrally hollowed in a trough-like manner; this part is hard and firm, shining, while the other part is more soft, lamellar, lying like a cover over the appendix-slit, and forward reaching to β. The terminal part on the ventral side is covered by a piece τ, quite corresponding to that in *Torpedo* and *Narcine*; the margins, however, are without folds.

The number of real terminal pieces is three.

Td is very small, a little concave on its ventral side; Tv is larger, and bears on its lateral edge a process directed forward; it is externally flatly rounded, internally towards the appendix-slit it has a trough-like concavity. T_3 articulates with the hindmost lateral edge of Rv, and with the fore end of Tv; proximally it is broadly ovate, and distally it tapers to a shining conical point, which, as already mentioned, is uncovered by the skin.

The muscular system is principally as in *Torpedo*; the *Musc. extensor*, however, is single, as in *Narcine*. The glandular bag occupies here, as in those two genera, the space between the distal end of the basale and the proximal end of the appendix-stem, but laterally it spreads over the hindmost rays in a similar manner as in *Raja*.

The glandular body is very narrow, and does not anteriorly reach the end of the bag; accordingly it fills the inner space of the bag to a far less extent than in the other Rays; it is prolonged as a thin, raised stripe, provided with a furrow and gland-pores as in *Torpedo* and *Narcine*, throughout the length of the shaft till the terminal part.

Trygonidæ.

Trygon violacea Bonap.

(Pl. III, fig. 38—40.)

A specimen of a length of 1^m, a breadth of 44^{cm} shows the following measures:

The length of the whole appendix . . . . 9^{cm}

 · — · · - free part . . . . 6,5^{cm}

from the fore end of the slit to the hindmost point of the appendix 8^{cm}

The length of the terminal part . . . . 4^{cm}

The largest breadth at the base of the terminal part. . . . . . . 1,5^{cm}

The whole appendix, as the ventral itself, is naked, somewhat latero-ventrally compressed, almost triangular, when cut through, but with rounded edges; the broadest surface looks inward, and is in contact with the base of the tail; the appendix-slit follows the more narrow dorsal side till the terminal part, where apparently it separates into two slits, surrounding a lengthened-oval, firm, and hard part, covered by the skin (*Tr*); in reality, however, only the inner one of these slits is a continuation of the appendix-slit (*af*); the outer one (*l*) leads into a deep pocket ending far forward, and limited by the terminal skeleton proper and the covering pieces to be mentioned later. Immediately before the passing of the slit into the terminal part its inner (dorsal) lip forms a rather large, soft, pigmented dermal fold, which, however, is not seen externally, being placed under the overlapping firm edge of the outer lip, supported by the covering piece *r*.

None of the skeletal parts protrude through the skin.

The skeleton. Between the short basale and the appendix-stem are found two pieces, b_1 and b_2, of about equal length, but b_1 is much the broader, especially proximally, where it is of the same breadth as the basale, and where on its dorsal edge it articulates with a very long piece β, which, without touching b_2, reaches to the appendix-stem, and articulates with the dorsal side of this almost quite to the fore end of the dorsal marginal cartilage. b_1 bears the last ray, which proximally for a long way is coalesced with the last but one which articulates with the end of the basale; these two are by ligaments firmly connected with the ventral marginal cartilage. The appendix-stem is long and powerful, more than twice as long as the basale + b_1 + b_2; its hindmost part (a little less than half the whole length) is uncalcified as a strong, broad style reaching to or only a little past the end of the terminal pieces. The marginal cartilages are both calcified, the dorsal one most solidly; they reach about equally far forward, and occupy almost the distal half of the stem above the style; behind the dorsal marginal cartilage reaches a little farther than the ventral one; the latter is concave on its outer surface, while the former is partly rounded.

Fig. 23.

Trygon violacea.
The left appendage from the dorsal side; reduced. *l* the opening of the lateral pocket. *l'*, *Td*, *Tr* the parts of the skin, in which the skeletal parts, indicated with the corresponding letters, are inclosed.

The number of terminal pieces is two, only *Td* and *Tr* being found.

Td is a large, externally rounded plate of pointed-ovate contour, thickest at the medial edge along the style, and laterally quite thin; the lateral edge is finely indented. *Tr* likewise is large, but the inner surface, towards the appendix-slit, is rounded, while the outer surface is deeply hollow as a trough, both edges of which are somewhat bent towards the concavity. The whole ventral side of the terminal skeleton, as well as great part of the ventral marginal cartilage, is covered by two hard, calcified, firmly connected covering pieces *r* and r_1, to which the *M. dilatator* is attached, and which correspond to the single *r* in the preceding genera of Rays (in *Narcine*, as we have seen, the proximal-lateral corner of *r* had already been partly separated as an independent piece, to which in *Trygon* the larger piece r_1 must be taken to correspond). The larger lateral piece r_1 shows distally a rather deep hollow on the outer, otherwise rounded surface, by which means a strong ridge is produced on the opposite, concave inner surface, and this ridge is firmly connected with the raised medial edge of the piece *Tr*; the above mentioned deep lateral pocket is then situated between the last-mentioned pieces.

The muscular system is peculiar by the little marked bounding between the single groups of muscles. Thus on the ventral side the fibres of the *M. adductor* are seen behind and medially to continue, without any bounding whatever, directly into the *M. dilatator*, and laterally to pass over on the *M. compressor* in such a way, that the contour of the glandular bag before and medially is quite effaced; the fibres of the ventral wall of the glandular bag — as in *Torpedo* — have the same direction as those of the *M. adductor*, and they are here, until close to the lateral edge, quite woven together with the latter. On the dorsal side the marginal portion of the *M. adductor* is seen as a powerful mass arising partly from the pelvis, partly from the aponeurotic covering of the muscles of the abdominal wall, and laterally overlapping part of the *M. extensor*. This latter arises from the whole extent of the basale, as well as from the following pieces, and is distally, without any bounding, woven together with the *M. adductor* and the continuation of this, the *M. dilatator*, which latter also with a considerable mass wraps the whole of the appendix-stem and both the marginal cartilages quite to their margins at the appendix-slit. Behind it is attached to the proximal edge of the ventral covering pieces (*v*, *v₁*). The part of the *M. compressor*, appearing as the outer lip-muscle, is only sligthly developed; more developed is the part of the dorsal muscular wall of the glandular bag, running more transversely, and wrapping the glandular body, which it follows throughont the shaft to its hindmost end at the terminal part.

The glandular body stretches here — as in the preceding genera of Rays — through the whole length of the shaft along the ventral marginal cartilage, constantly tapering backward, and the furrow which in the part of the body, situated in the bag, runs from before obliquely to the medial side (separating a pigmented and an unpigmented surface), continues with its pores quite to the terminal part. The part of the appendix-slit situated in the terminal part, has quite smooth walls.

The above described features of the glandular body, and several other characteristics — as for inst. the presence of ventral covering pieces, enclosing a pocket with an opening through an outer side-slit, and furthermore the hollowing of the ventral side of the piece *Tr* — very much recall those in *Torpedo* and *Rhinobatus*[1]), and it is evident that *Trygon* is more closely allied to those than to *Raja*, at all events with regard to the structure of the appendices.

Rajidæ.

In this family the appendices reach to quite a considerable size, and consequently they, naturally enough, have attracted the attention, and have in some species several times been the subject of examination. As a rule, however, this examination has not been very thorough, nor has it been ex-

---

[1]) Partly even those in *Rhina*.

tended to more species of the genus, and therefore the result as to comparison has only been slight. The species show great differences that may easily be used diagnostically. The corresponding structures in the different species are easily pointed out; more difficult it is to work out the comparison with the other Plagiostomes in a sure way. I hope, however, to have succeeded in this in the following description of the species I have examined.

Raja batis L.
(Pl. IV. fig. 45—48.)

The appendices of the Skate are mentioned by several authors, who, however, have restricted themselves to brief remarks of the outer shape and the size. This latter may in old males be so considerable that the appendices may convey a notion of Skates with three tails , or Skate-Kings (Pontoppidan)[1]; Lilljeborg[2] gives, in old males, the length to be between $^1/_4$ and $^1/_5$ of the total length of the animal, and says that the appendages reach far behind the middle of the tail. I know, however, no thorough representation of these organs or their skeleton[3].

In a specimen of a length of 1^m 26^{cm} (2 ell Danish) the appendix has a length of $27,5^{cm}$[4]. The other measures were:

From the fore edge of the slit to the end of the appendix   25—26^{cm}
The part free of the ventral . . . .   18^{cm}
The terminal part . . . .   13^{cm}
The largest breadth (across the basis of the terminal part)   4^{cm}

The shape of the appendix is flattened, the contour clavate, the breadth increasing towards, and culminating in, the big, ovate terminal part. The skin is naked in every place. The appendix-slit in the free part of the organ is situated quite close to the lateral edge; in the part united with the fin more towards the middle of the dorsal side; in front the slit is easily dilated, and a little-finger may here be brought into the ventral glandular bag, the powerful gland of which may be partly discerned through the skin; the rest of the slit until the terminal part is certainly open, but on account of the stiffness of the marginal cartilages it can only be very little dilated; in the terminal part, however, dilation may easily take place, especially if the end of the appendix is bent ventro-medially. In the terminal part the skin forms on the ventral side a large, soft lip, closing together with the dorsal lip, which is supported by skeletal parts. If the soft lip is thrown back, its inner, bluish-red mucous membrane is seen, as also a large skeletal piece (T_3) with sharp, indented outer edge; it is, however, quite covered by the mucous membrane, which on the dorsal side of the piece

[1] See Kroyer, l. c. p. 993.
[2] l. c. p. 590.
[3] The Ray mentioned by Joannes Battarra in Atti dell' Acc. delle scienze di Siena, Tomo IV, 1771, p. 553, the appendix-skeleton of which he draws in fig. I, must be the Skate, or at all events a nearly allied species (according to Giglioli [teste Lilljeborg] the Skate is not found on the coasts of Italy(?)). Davy l. c. p. 145 mentions the glandular bag and its large glandular body, its secretion etc.
[4] In two skeletons of ventrals belonging to the Zoological Museum the appendices have a respective length of 41^{cm} and 38^{cm}.

forms a great many transverse, soft, vascular folds (see fig. 2) in the text, *b'*). If the terminal part is more opened, the walls of the appendix-slit are seen; they are very curiously formed, covered with a mucous membrane, and their folds and pockets supported by different skeletal pieces. Besides what has been mentioned, we find on the ventral side a firm fold with a porous edge, *da*, reaching into the proximal part of the slit, and a process (*Tr₂*), with a loose, soft covering, and supported by skeleton; laterally of this process is found behind a deep recess or pocket, *Lr*. On the dorsal side are seen two recesses, the foremost one (*Ld*) very deep, and separated from the appendix-slit by a lamella supported by skeleton; the hindmost one (*Ld'*) is less deep; finally is seen a process (*Tr*), enclosed in a soft membrane, which process is laid against the above mentioned one on the ventral side of the slit. The real continuation of the appendix-slit runs between these two processes, as indicated by the sound in fig. 2.) ¹).

The skeleton. Between the basale and the appendix-stem are found two pieces, b_1, b_2, the latter longer than the former; b_1 bears the six hindmost rays, b_2, as usual, none; with the dorsal and lateral edge of b_1 is connected a long, plate-shaped β, distally articulating with the appendix-stem *b* almost beside the articulation between this latter and b_2.

The appendix-stem is long, about twice as long as the basale + b_1 + b_2; behind it becomes by and by dorso-ventrally flattened, especially in the terminal part, where its outer end is quite flattened, thin, and rounded. As is usually the case, the calcification ceases in the terminal part; in the long part corresponding to the style is however found on the medial edge a strongly calcified region projecting in a somewhat bump-like manner (*x* in fig. 45, 47). The marginal cartilages are long and hard, and are for a long way rather closely joined with their edges; distally they separate; the dorsal one begins before close by the articular surface between β and *b*, but distally it does

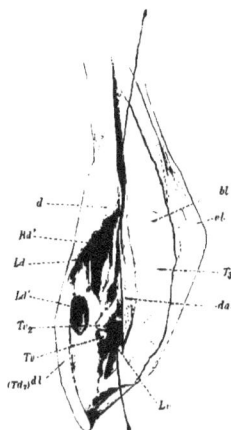

Fig. 2.). *Raja batis.* The terminal part of the right appendage, from the dorsal side, strongly dilated (much reduced). A sound goes through the bottom of the appendix-slit. *v'* the soft, ventral dermal lip; T_3 the terminal piece T_3 covered by mucous membrane with the dermal leaves *bl*; *da* a firm dermal fold, dotted with small holes on the edge. *Lr* the ventral pocket, *Ld*, *Ld'* the two dorsal pockets; *d*, *d'* the dorsal lip with the enclosed skeletal parts *d* and *Td₂*; *Rd'* a dermal fold, supported by a plate-shaped prolongation of the dorsal marginal cartilage; *Tr*, *Tr₂* dermal projections supported by the skeletal parts marked with the same letters.

not reach as far out on the stem as the ventral one, which in return does not reach so far proximally as the dorsal one; thus an open space is found proximally, where the glandular bag joins in, and where its foremost, dilatable outlet is situated. The dorsal marginal cartilage sends forth distally a

¹) In specimens with quite short, undeveloped appendices, shorter than, or of an equal length with, the fin-membrane (the free part from 2.5 — 3.5ᶜᵐ in length), we, as might be expected, do not find much of these elaborate structures; in such young appendices with all parts still soft the appendix-slit is easily opened and quite spread, and then the walls of the slit are seen to be smooth and simple, upon the whole without recesses or folds, also in the terminal part; one strongly pronounced dermal fold is however seen along the ventral lip; the inner, already plate-shaped edge of this fold represents the spongy-porous dermal fold (*da*) in the adult, and in the lateral part of it the large skeletal part T_3 will develop.

long, triangular, plate-shaped prolongation, Rd, fig. 45, running into the terminal part, and supporting the above-mentioned dermal fold.

The number of terminal pieces is five, and to these is to be counted a «covering piece» belonging to the aponeurosis of the *Musc. dilatator*. This covering piece[1] (fig. 48) is chiefly situated on the dorsal side, has a fairly triangular contour, flatly rounded, with part of the medial edge bent in such a manner that it catches round the medial edge of the end-style (fig. 46, d). This piece quite evidently belongs to the same kind of structures as those, described as covering pieces in the before mentioned forms (*Rhina*, *Mustelus antarcticus*, *Torpedo*, *Narcine*, *Trygon*), only the development having here taken place in the dorsal side of the aponeurosis; and so I here (and in other *Raja*-species) mark it with a d.

With the distal end of the dorsal marginal cartilage is only connected one piece, proximally also touching the style; I regard it to be corresponding to the Td of the other Plagiostomes, as it is situated in the same manner as that in the dorsal lip of the appendix-slit, and shows the same relations to the marginal cartilage and the stem; it is rather thick, somewhat bent in a crescent-shaped manner with the concavity towards the stem; distally it is firmly connected with the style of the stem by means of an uncalcified (or slightly calcified) cartilaginous prolongation[2]. With the hinder half of the lateral edge of this piece Td is connected a long, narrow piece, attached by a band to the terminal end of the style; it is situated in the edge of the dorsal lip, and I think it to be corresponding to the piece, which in several other Plagiostomes I have called Td_1. With the ventral marginal cartilage are terminally and laterally connected two pieces of a peculiar shape. Only the medial one of these is also connected with the style of the stem, passing closely along it until the above mentioned calcified bulb x; this piece then must be corresponding to Tv; it is rather thin, rounded on the outer side, and prolonged to a slender, bent, obliquely-posteriorly and laterally directed part, ending in a hook (it is this hook, which is seen covered by a loose, mucous membrane in the fig. 24 of the text at Tv). The other piece consists of a crescent-shaped, large part, the concave side of which is for a long way connected with the terminal and lateral edge of the ventral marginal cartilage; most anteriorly the long horn of the crescent reaches into the appendix-slit; behind and laterally the piece is prolonged to a long, rather slender process, tapering to a somewhat bent, flat part with a slightly notched end (it is this process, which is seen covered by loose, soft membrane in the fig. 24 of the text at Tv_1). I take this piece to be homologous with the fully developed piece Tv, in *Spinax* (in *Acanthias* and *Somniosus* it is only indicated), and its situation seems to me to settle this homology beyond doubt. Finally is found, belonging to the ventral structures, a very large piece, rather crescent-shaped when viewed from the ventral side, which must be corresponding to the piece T_3 (the thorn or spur in different Sharks); proximally it reaches far into the appendix-slit, is on its medial side broad, flatly rounded, and sends forth laterally a high, sharply winglike ridge with undulating, finely indented edge; somewhat above the middle it sends forth a tap-like process, ventrally overlapping Tv_1; distally it ends with a plate resembling the blade of an axe, and ventrally reaching over the

---

[1] It is mentioned by the way by Morena l. c. p. 251 under *Raja clavata* by the name of Cartilage interne»; the presence of Td_1 is also noted here, but this piece gets no special name, as it is not found in *R. clavata*.

[2] This connection-piece I take to be not independent.

style[1] (to judge by the pieces before me this part seems to be a little varying with regard to the details of the processes).

As to the muscular system the reader is referred to *Raja clavata* with which the other *Raja*-species agree exactly in most respects. The *M. dilatator*, however, is here somewhat more distinctly than in the thorn-back separated into two, a large dorsal one, and a smaller ventral one.

The glandular body, as in all *Raja*-species, is limited to the dorsal side of the ventral bag proper, and accordingly does not continue into the shaft. In young specimens of the Skate with undeveloped appendages not yet reaching the end of the ventral, the appearance of the gland is rather deviating from that of the developed one; it is more flattened, and the gland-pores are spread in more (four or more) rows over the greater part of the surface; in other words, it shows a rather striking resemblance to the glandular body in *Rhina*. By a continued growth of the marginal portions of the gland, and by a strong rounding of these portions, the surface that in younger animals is provided with pores, will be hollowed, and thus apparently become more narrow, and in this way the characteristic deep longitudinal furrow will arise.

Raja nidarosiensis Collett.

Of this species I have had only one pair of dried ventral skeletons, the appendix-part of which had a length of abt. 33^{cm}. The principal features agree exactly with those of the Skate; the resemblance includes the common habit of the single skeletal pieces, but a closer examination of these will show some minor peculiarities in the details; for inst. it will especially easily be seen that the piece T_3 is bent in a somewhat different way, and has a relatively larger lateral wing, a larger forward directed tap, and its distal end has not the peculiar shape like the blade of an axe; on its medial surface it is deeply concave, spoon-like, etc. I shall however omit to give a detailed account of the deviations of all the pieces from those in the Skate, and leave to others, who may have more material at their disposal, to work out this subject more thoroughly; no doubt the deviations will make good specific characters. According to the existing descriptions of the species the appendages are very large, and are said to reach behind to the beginning of the first dorsal fin.

Raja clavata L.
(Pl. IV, fig. 49–52; pl. VI, fig. 67–68.)

This Selachian I think to be the one whose appendices have most frequently been examined and described. Duvernoy[2] has briefly mentioned them, their glandular bag with its glandular body and their skeleton, the nomenclature of which he has formed in accordance with the appellations of the parts of the hind limb and the foot of a mammal. Later they have been described by Vogt and Pappenheim[3], whose description it is rather difficult to use on account of the want of references

---

[1] In my figure it has been displaced a little, so as to reach too far up on the other terminal part.
[2] Cuvier: Leçons de l'anatomie comparée, 2d ed., vol. 8, 1846, p. 305 (perhaps the description applies to *R. circularis*?).
[3] Reich. sur l'anat. comp. des organes de la génération chez les animaux vertébrés. Ann. sc. nat. (Zool.), vol. XII, 1859, p. 111–117, pl. 3.

to the accompanying figures, and in these the total want of letters indicating the single parts; still later they have been described by Petri[1]) and Moreau[2]). Besides short remarks on these organs are found in several writers of mostly systematic works. None of these descriptions seem to me to be quite serviceable.

In a specimen of the total length of 76^{cm} the whole length of the appendix is 18^{cm}. The other measures are:

Length of the free part . . . .   13^{cm}
from the fore end of the slit to the extremity of the appendix   15^{cm}
from the fore end of the terminal part . . . .   10^{cm}
Largest breadth (across the base of the terminal part). .   2,7^{cm}

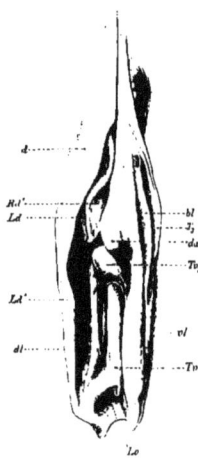

Fig. 25.

Raja clavata. The terminal part of the right ventral appendage, much dilated; reduced. The letters as in fig. 24.

The whole appendix is naked (but on the abdominal side of the glandular bag are found some scattered thorns); the shape is lengthened-clavate with a short shaft; the terminal part forms the club, and is relatively very long, broadest at the base, and tapering from thence towards the point. Only the foremost dilatable part of the appendix-slit, the part forming the foremost outlet from the glandular bag, is situated dorsally; from here the slit goes laterally, almost even passing to the ventral side; from this side it can be seen, but not at all from the dorsal side (contrary to the situation in the Skate). This is brought about by the fact that the dorsal lip, which only proximally is supported by skeletal parts, at the base of the terminal part overlaps the slit to such a high degree; the ventral soft lip is, when compared to that of the Skate, only very narrow. If the terminal part is opened so much, that the interior is seen[3], this latter will present an appearance, apparently quite different from that in the Skate; it will, however, be possible to point out quite corresponding projections and hollows: on the ventral side of the slit is seen, relatively only little conspicuous, the membrane-covered terminal piece T_3[4] which shows before a sharp, cutting edge, and on closer examination also is seen, as in *Raja batis*, to bear on its upper surface a row of transverse, soft, but less developed dermal leaves *bl*; a bayonet-like, hard, and sharp-edged blade[5] (on which one may easily cut one-self, although it is covered by membrane) projects strongly, corresponding to Tr_2 in the Skate; to the piece Tr in this latter corresponds a

[1]) l. c. p. 310.
[2]) Hist. nat. des Poissons de la France. vol. I, 1881, p. 248—259.
[3]) Petri, l. c. pl. XVI, fig. 1 D, has given a drawing of the dilated terminal part, which is quite unsatisfactory, especially with regard to the ventral side.
[4]) Petri, l. c. fig. 1 D, *hk.*
[5]) Petri, *bf.*

long cylindrical process, reaching with the point into the ventral recess[1], which is here situated far backward, almost terminally (Lv); on the dorsal side are seen the two same recesses[2] as in the Skate, but of different size and extent, as well as the projecting lamella Rd'[3] borne by the skeleton; the only new thing that does not seem to be represented in the Skate, is a strongly projecting part supported by the skeleton, da[4]), which part belongs to the ventral side; it corresponds to the porous, spongy, firm dermal fold in the Skate, where, however, no skeleton is found.

The Skeleton. Between the basale and the appendix-stem (Metatarsien Moreau [Cuv., Duvernoy]) there is a b_1 bearing the last six rays, a b_2 without rays, a flat β with s-shaped edges, proximally connected with b_1, and distally by means of a joint with the appendix-stem[5]). This latter is about (scarcely) twice as long as $B - b_1 + b_2$, and closely behind the two proximal articular surfaces for b_2 and β it becomes flattened, and ends with a quite flat, thin-edged, hindmost rounded part; corresponding to the great length of the terminal part, the uncalcified style (une sorte de phalange Moreau [etc.]) forms the greater part (about two thirds; in my figure it is a little too short); also here is found, about at the middle of the medial edge, a thickened and somewhat calcified part. The marginal cartilages are large, plate-shaped; the dorsal one reaches farthest forward, but behind it ceases much before the ventral one; the dorsal cartilage sends forth into the interspace between the terminal pieces a long, firmly calcified, s-shaped and pointed, blade-like process, bent in an undulating way, and with sharp edges (Rd', fig. 50).

The terminal pieces are 5 (6), to which is still to be counted a covering piece[6]). This (d) is situated dorsally, is flatly rounded on the outer side, concave towards the terminal part, whose proximal portion it covers; anteriorly it is curved in an oblique, half-moon-shaped manner, having the lateral fore corner far drawn out; from this corner it shows an elevation running towards the medial edge, and indicating the place of attachment of the $M.$ $dilatator$; consequently the whole part before this line is covered by the muscle. Td[7] has a broad line of attachment with the distal edge of the dorsal marginal cartilage, but only touches the appendix-stem; outward it is somewhat flat, inward concave; its lateral edge is convex, and somewhat indented; its distal end is attached to the appendix-style by soft tissue, representing the cartilaginous bridge in the Skate. The piece Td_2 found in $R.$ $batis$ is wanting here. Tv is a long, slightly bent piece[8]), provided with a short, hook-like point, and having before a short articulation with the distal end of the ventral marginal cartilage, medially a long connection with the appendix-style, reaching until the thickened and calcified place in this latter; laterally it is for a rather long way connected with a bayonet-like Tv_2[9]); this latter is connected with the hindmost lateral edge of the ventral marginal cartilage by an oblique articulation, and distally it

[1] Petri $etc.$
[2] Petri $ala, alp.$
[3] Petri $pr.$
[4] Petri $da.$
[5] Petri, fig. 1 C, b_1, b_2, β'; Moreau (= Cuvier-Duvernoy): Tibia, astragale, calcanéum; basale = fémur.
[6] Petri: sch' schuppenförmiger Knorpel , grössere Schuppenlamelle ; Moreau: Cartilage interne, no. 3, fig. 27, l. c. p. 250.
[7] Petri: sch, likewise schuppenf. Knorpel , kleinere Schuppenlamelle (p. 312); Moreau: Cartilage externe, no. 2.
[8] Petri: st; Moreau: Cartilage no. 6 bis, cartilage en cuilleron.
[9] Petri: bj; ein bajonettähnliches, gedrehtes Knorpelstück (p. 313); Moreau: Cartilage en hallebarde (Duvernoy), no. 6.

runs out in an elegantly shaped, longitudinally somewhat twisted blade with a peculiar sharp lateral edge. Ventrally of this, and attached to the same edge of the ventral marginal cartilage, is found a piece *da*[1]) which I do not find in the Skate, or in any of the other Plagiostomes I have examined, but, according to Petri and Moreau, it evidently appears in several other *Raja*-species; it has a thick, lateral edge, and a rounded contour; it is movable and seems to be composed of two pieces, a little, proximal, lamellar *du'*, and the larger distal *da*. Finally is found the large piece T_3[2]). It consists of a more narrow foremost part, the proximal end of which is attached to the lateral edge of the ventral marginal cartilage, and which laterally sends forth a sharp, winglike ridge (corresponding to that in the Skate, but much lower), and next of a broader, hindmost part whose medial edge (corresponding to the axe-blade in the Skate) folds round the appendix-style[3]).

The muscular system, with regard to the proximal part, shows the typical relations, as will be sufficiently clear from the figures 67 and 68 on pl. VI.

The *M. dilatator* is on the dorsal side behind split into two parts, but this cleaving has not been carried through to the proximal part of the muscle, and so the *M. dilatator* seems to me to form one muscular mass here as well as in most of the other Plagiostomes I have examined. The whole of the large dorsal part of this muscle is with its hinder end attached to the dorsal covering piece, i. e. not to the edge of this piece, but some way in on its surface till a plainly indicated line of insertion (see fig. 49 ou pl. IV). In *Raja batis* the division of the *M. dilatator* indicated in *R. clavata* seems to be more strongly pronounced, and in other *Raja*-species[4]) it even seems to lead to a separation into two independent muscles, one larger situated dorsally, and another smaller, ventral, which

<hr/>

[1]) Petri: *da* ·ein spatelförmiger Knorpelstück (p. 313); Moreau, who has correctly seen that it is composed of two pieces, calls the little proximal one: Cartilage intermédiaire, no. 4, the larger one: Cart. accessoire, no. 3.

[2]) Petri: *hk* hakenförmiger Knorpel ; Moreau: (Duvernoy) Cartilage en soc de charrue, no. 7 (in the principal figure, however, indicated by *t*).

[3]) Vogt & Pappenheim's appellations have to be with difficulty found out from the description, this, as mentioned, having no references at all to the figures, and in these no letters are found. I give below the appellations of these authors corresponding to my names. It seems that they have not clearly seen that the chief piece -- la pièce principale -- is composed of three parts; they use the names ·la lèvre interne · partly of the marginal cartilages, but without establishing the independence of these pieces; the prolongation of the dorsal marginal cartilage is described (p. 114) as ·une feuille mince en forme de spatule·. The other names are:

The covering piece *d* = pièce externe, la plus superficielle (p. 115).

Td = pièce externe; seconde pièce.

Td' = pièce alongée, courbée en S; it is interpreted as coalesced of two pieces, the terminal part wrapping the appendix-stem being called ·une petite pièce cartilagineuse formant une gouttière· etc. (this part in young animals is possibly soft).

da = petite pièce cartilagineuse .... presque carrée et couverte .... par un coussin gélatineux.

*Tv*3 = pièce .... plus allongée, sa forme est semblable a celle d'une équerre très large.

Tv = une dernière pièce cylindrique etc. (p. 116).

[4]) For inst. in *R. Schultzii*, according to Petri (l. c. p. 314; pl. XVII, fig. 2 B and C). Petri calls the greater, medio-dorsal part *M. levator*, and thinks this part to be composed of two kinds of muscles, viz. the greater part of red fibres in which is found a wedge-shaped · white· part (fig. 2 B, *al*) the fibres of which, however, are said to run — only with altered colour – into the red mass (a difference of this nature I never saw in any Plagiostome): the smaller, dorsal muscle is called *M. rotator* with regard to its action on the hakenförmige Knorpel (my piece *T₃*). In *Raja clavata* Vogt & Pappenheim (l. c. p. 116), as it would seem, (the description is not quite clear to me) have also found two muscles where I only find one; they speak of a ·Muscle écarteur dorsal·, originating · on the large dorsal covering piece [where the fibres of their · M. releveur· are said to be attached; according to their description this releveur (or one thing is composed of the dorsal layer coming from the body (i. e. the tail), and is rather incomprehensible to me]; and next of a ·muscle écarteur ventral· which by means of rather long tendons· is attached to the outer side of the ·s-shaped piece· (*T*). Duvernoy also says (l. c. p. 308) that his ·Muscle grand abducteur· (*M. dilatator*) in ·la raie ronce· is divided in a similar manner, and attached in the same way.

latter during the dilation acts especially on the terminal piece T_3, by which means this piece is turned (revolving as the radius round the ulna in a human fore arm).

The $M.$ $compressor$ (S) forms the bag situated on the ventral side; the direction of its fibres, as far as seen on the ventral side, is exactly given on pl. VI, fig. 68; on the dorsal side (fig. 67) the part forming the outer lip-muscle, which part is rather small, is seen anteriorly arising from the piece β, and posteriorly attaching itself in the inner investment of the ventral marginal cartilage; when the connective tissue laterally uniting it with the hindmost ray, is prepared away, it is here very distinctly seen to be continuously connected with the dorsal muscular wall of the bag [1].

The glandular body has by earlier authors been sufficiently described as well in this species as in other $Raja$-species; when developed it seems in all species to show principally the same appearance.

Raja radiata Donovan.

(Pl. IV, fig. 53—57.)

Brief remarks on the appendices of this species are found in several authors, as usually mostly concerning the size [2]) and the like facts. Lilljeborg [3]), however, not only says that they are very large and in old individuals sometimes reach past the middle of the tail (in a specimen of the length of 53^{cm} they were 14^{cm} long and 3.2^{cm} broad on the middle), but he also gives a rather thorough description of their outer contour and whole shape; of the inner configuration of the appendix-slit he only says that it is "divided into parts or separate hollows". He does not enter upon an examination of the parts of the skeleton; he mentions only, that a piece in the dorsal lip has a free, backward directed point. We find, however, in the older literature a representation of the skeletal parts of these organs, as well as of their structure upon the whole, viz. by M. E. Bloch [4]). His specimen had a length of 16 inch. (about 42^{cm}), and the appendices (from the pelvis) were 5$^1/_2$ inch. (abt 15^{cm}) long, 1$^1/_8$ inch. (abt 3.5^{cm}) broad across the terminal part.

[1]) Petri has called this part of my $M.$ $compressor$ $M.$ $flexor$ $pterygopodii$ $biceps$ (l. c. fig. 2 B, fb), and thereby indicated that he thinks it to be corresponding to the muscle in $Scyllium$ marked with the same name, which latter, however, shows quite other relations (see my fig. 65 and 66 of $Sc.$ $stellare$); he describes it as inserting itself on the dorsal marginal cartilage instead of on the ventral one (this, perhaps, is only a miswriting). In the proximal part the $Flexor$ $pterygopodii$ $exterior$ of Petri corresponds to my $M.$ $adductor$ (A), his $Flexor$ $pt.$ $interior$ to my $M.$ $extensor$ (E). The muscles mentioned by Duvernoy (in Cuvier's Leçons 2d ed. vol. 8, p. 307) are: 1) Le muscle abaisseur = my $M.$ $adductor$; 3) L'abducteur de l'appendice = $M.$ $extensor$; 5) Le grand abducteur ou extenseur des pièces mobiles et terminales = $M.$ $dilatator$; his no. 2 le releveur de la nageoire is the muscular layer coming from the body (pl. VI, fig. 67), and his no. 4 (Morean's court extenseur) I am unable to unravel. The same names have mostly been used by Moreau (l. c. p. 255); his Muscle long extenseur = $M.$ $extensor$, his M. fléchisseur = the part of $M.$ $compressor$ forming the outer lip-muscle. Moreau, in correspondence with my opinion, describes his M. grand abducteur ($M.$ $dilatator$) as separating into two bundles. Vogt & Pappenheim, besides the already mentioned (écarteurs) ($M.$ $dilatator$) only mention the $M.$ $adductor$ as l'abaisseur de l'appareil copulateur , and as the antagonist of this a releveur partly formed by the dorsal muscular layer coming from the body. Bloch only mentions two muscles in $Raja$ $radiata$ (l. c. vol. 6, 1785) both together representing my $M.$ $dilatator$.

[2]) Kroyer l. c. p. 943 gives the measures: a specimen 17$^1/_2$ inch. long with appendices of the length of 4$^1/_2$ inch., and thinks (p. 954) that the appendages are very strongly developed in the adult males.

[3]) l. c. p. 552.

[4]) Von den vermeinten doppelten Zeugungsgliedern der Rochen und Haie. Schr. d. Berl. Ges. naturf. Freunde, vol. 6, 1785, p. 377. Bloch calls his Ray-species $Raja$ $clavata$ L., and in his faunal works he has drawn and described it as $Raja$ $clavata$. The figures, l. c. pl. IX, however, show with perfect certainty that the species in question is $R.$ $radiata$. (Both this plate and the one concerning $Acanthias$ are, without any explanation, affixed to the edition by Schneider of Bloch's Ichthyology.) Petri has not perceived that in Bloch the question is not of the real $R.$ $clavata$.

Two pairs of ventrals with fully developed appendices, now before me (unfortunately I cannot give the total length of the animals) show the following measures:

Length of the appendix . . . . 10^{cm}
- - part free of the fin . . . . 6,2^{cm}
- - terminal part . . . . 4,6 –5^{cm}
Breadth of the   —   —   . . . . 2,7^{cm} at the broadest spot
.— - - basal part . . . . 1,6^{cm}

A dried specimen of a length of 39^{cm}, a breadth of 24^{cm} shows fully developed appendices, 8^{cm} long, and 2,2^{cm} broad across the terminal parts; in another specimen (in spirit) 43^{cm} long, and 29^{cm} broad the appendices are only 6,5^{cm} long, the terminal part abt. 4^{cm} long, 2,1^{cm} broad; here they are not yet fully developed though it was to be expected judging from the size of the animal. Facts as these, that rather grown individuals have rather undeveloped appendices, I have oftener seen, for inst. in *Acanthias*.

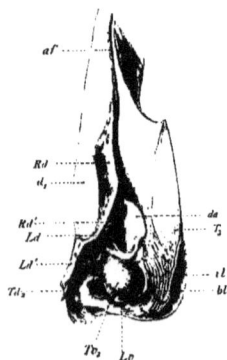

af
Rd
d,
Rd'
Ld
Ld'
Td,
da
T,
il
bl
Tv,
Lp

Fig. 26.

Raja radiata. The terminal part of the right ventral, much dilated; reduced. *Td,* the naked spine of this same skeletal piece. The letters as in fig. 24 and 25.

The appendix is naked, much more clumsy than in the preceding species, flattened, somewhat rounded on the dorsal side, the contour is clumsily clubshaped; the club is formed by the terminal part constituting more than half the length of the part to be seen from the back. The appendix-slit runs from the foremost dorsal opening laterally, so that it cannot be seen from the dorsal side except in the hindmost end of the terminal part, where the dorsal lip, as it were, retires; the dorsal lip, throughout the terminal part, is supported by inner skeletal parts reaching to its edge, while the soft membrane of the ventral lip as a broad wall stretches past its skeletal part (*T,*), and is laid – in a similar manner as in the Skate — dorsally against the upper lip; from the hinder, lateral edge of this latter a naked spine projects. If the soft, ventral dermal lip *rl* is thrown back, an elevated, long, bowshaped, cutting edge of the skeletal piece *T,* is laid bare (fig. 26 in the text, to the right of *bl*). If the terminal parts are opened still more (which here is easily done), we shall, although with altered shapes and relations, see corresponding projections and hollows as those described in *R. batis* and *clavata*. The upper side of the piece *T,* does not here show (or shows at most weak traces of) the transverse folds *bl*, peculiar in those two species; a broadly tongue-shaped, rather soft and movable lamella with porous edge and spongy lateral surface represents *da* in the Skate and the Thorn-back; a large, ovate, hard swelling corresponds to the process *Tv,*; behind and laterally of this the ventral recess *Lv* is found, large and deep; the foremost recess *Ld* is smaller and more hidden, situated before the ovate swelling, and also the lamella *Rd'* supporting its lateral wall, is only little conspicuous.

The skeleton. Between the basale[1]) and the appendix-stem, as in all *Raja*-species, are found a b_1, a b_2, and a β^2); b_1 also here bears the last six rays. The appendix-stem is about twice as long as $B + b_1 + b_2$, flattened especially distally; its terminal part, being as usual uncalcified, here forms an s-shaped, quite thin, flat style broadening towards the end. As in the other *Raja*-species the dorsal marginal cartilage stretches forward almost to the beginning of the stem, but backward not so far as the ventral one; this latter, especially distally, is a good deal broader. To the inner side of the dorsal marginal cartilage, at its distal edge, is articulated a triangular cartilage *Rd'* (fig. 56); it is quite corresponding to the one marked *Rd'* in the two other *Raja*-species, in which, however, it is only a direct prolongation, a process, from the marginal cartilage itself.

The number of terminal pieces is five, exclusive of the covering pieces. Three covering pieces d_1, d_2, d_3 are found on the dorsal side. The lateral one, d_1, is a good-sized, externally rounded plate, with a bow-shaped, convex lateral edge folding round to the ventral side. Its medial edge is rather straight and firmly connected with d_2, which latter as a narrow band runs obliquely across the terminal part, and tapers towards the medial end that is bent round to the ventral side, and by a ligament attached to the point of the piece T_3 (see fig. 55). The third covering piece, d_3, is connected with the lateral end of the preceding one; it is of a triangular, externally somewhat rounded shape, and by a ligament attached to the hindmost end of the appendix-style; with its inner surface is is connected with the dorsal side of the piece Td_2. The three mentioned covering pieces have all arisen from the same aponeurosis of the *M. dilatator*, and accordingly they together represent the single covering piece *d* in the Skate and the Thorn-back.

Of real terminal pieces two are found in the dorsal lip: *Td* and Td_2 (see fig. 54, 55). *Td* is short; it is with its whole fore edge attached to the dorsal marginal cartilage, with its foremost medial corner also to the appendix-stem; from its medial-distal corner it sends forth a soft, cartilaginous part which farther backward is coalesced with the style (comp. the Skate); else its distal edge is connected with Td_2, a proximally broad, distally narrow and tapering, very hard, somewhat s-shaped cartilage; it is outwardly rounded, inwardly concave, and ends in the above mentioned thorn projecting naked from under the edge of the dorsal lip.

To the ventral side belong three terminal pieces: Tv, Tv_2, and T_3. The two first of these are very peculiar, and can only be rightly seen when the skeletal parts are disunited (see fig. 57).

Tv consists of two parts, a body and a long process; the body is proximally attached to the edge of the ventral marginal cartilage, with one edge to the medial edge of the style (see fig. 54), and with the opposite one to the piece Tv_2; from the ventral surface of the body the process arises, and forms together with the body a kind of T; this process is bent in an irregularly s-shaped manner, ends in a fine, hook-shaped thorn, and is situated in the deep, spoonlike hollow formed by the piece Tv_2.

Tv_2 is still more peculiar; its chief part forms an oval spoon, outwardly strongly rounded, inwardly very deeply hollowed, from the foremost part of which a large, half-moon-shaped part arises joining the inside of the ventral marginal cartilage of the appendix-stem; the lateral edge of the spoon is prolonged into a not quite calcified, winglike process; between this process, the half-moon-shaped

[1]) Bloch l. c.; der erste Knochen des Schenkels , fig. 1, *f*.
[2]) $b_1 =$ der zweite , $b_2 =$ der dritte , $g =$ der vierte Knoch, u der Schenkels , fig. 1, *h*, *u*, *o* in Bloch.

part, and the firm body of the spoon part of the terminal piece T_3 is intercalated as a kind of articular head (see fig. 55).

This latter, T_3, is large, forms a half-moon-shaped plate (fig. 55), the distal horn of which is bent in a somewhat hook-shaped manner; the proximal horn stretches in between the marginal cartilages, far forward in the appendix-slit; on the concave edge of the half-moon the mentioned articular head[1]) projects bearing a large, transverse-oblong articular surface ; the upper or inner surface (the surface towards the appendix-slit) bears at the lateral, convex edge a thin, bent, sharp ridge which in some individuals is undulating or finely indented; it is the above mentioned edge seen on the undamaged organ[2]).

Still has to be mentioned a peculiarly elevated, round, narrow cartilaginous ridge x, running across the dorsal surface of the appendix-style; this ridge seems to me only to be a special swelling of the style, and to correspond to the calcified and thickened bump x in $R.$ batis and clavata.

Raja fyllæ Ltk.

In a specimen[3]) of a total length of 55^{cm}, a breadth of 30.5^{cm}, the fully developed appendices are 11^{cm} long, i. e. exactly $\frac{1}{5}$ of the total length.

The other measures were:

From the beginning of the slit to the end of the appendix . . . . 9^{cm}
The part free of the fin . . . . 7.6^{cm}
The length of the terminal part . . . . 5.5^{cm}
The breadth across the shaft . . . . 1.5^{cm}
　—　　—　　the terminal part . . . . 1.75^{cm}

As in the other Rajæ the appendix is naked. The outer form as well as the inner configuration of the appendix-slit in the terminal part is very much like that of the Skate. The contour consequently is of a more slender club-shape than in $R.$ clavata or radiata with a longer shaft and a pointed-ovate, somewhat broader club constituting the larger, hinder portion of the terminal part. As in the Skate the appendix-slit can be seen for its whole length from the dorsal side, but runs close

---

[1]) This evidently corresponds to the tap on the piece T_3 in the Skate, which overlaps the piece Tr_2.

[2]) Bloch l. c. pl. IX has drawn most of these terminal pieces in a very recognizable manner, some of them even excellently (as fig. 4 and fig. 5). He distinguishes between an upper part (the chief piece of the appendix), and a nether part (the terminal part); the first he interprets as a tibia with its fibula (?) (ein Röhrenknochen, und sitzet letzterer oberwärts, wie bey anderen Thieren, an dem Schienbein fest); this latter is = my dorsal marginal cartilage; l. c. fig. 1 and 3, q; the tibia again consists of: a piece (= my ventral marginal cartilage), l. c. fig. 1 and 3, $r, r,$ welcher unter gewissen Umständen die Rinne verschliesst , and of an unterer Knorpel (= my appendix-stem), fig. 3, s; it ends hooklike; this is brought about by the fact that Bloch has not separated the covering piece d_1 from its connection with the style. Bloch makes the nether part consist of five pieces, which number arises from the fact that twice he makes two pieces one. These five pieces have the following relations to my appellations:

T_1 = fig. 4, der Sichel .
Tr_2 = - 5, der Helm .
Tr = - 6, der wurmförmige Knochen .
$Td - Td_1$ = - 7, der Winkelhaken .
$d_1 + d_2$ = - 8, die Schaufel .

[3]) Station 25, at a depth of 582 fathoms; the Davis Strait.

to the lateral edge; the dorsal lip of the terminal part, also like that of the Skate, is along the whole lateral edge supported by skeletal parts, while the ventral lip has a broad, soft edge covering the skeletal parts (T_3), and passing round to the dorsal side; if this dermal lip is thrown back, we shall see, quite as in *R. batis*, a naked, cutting edge of a raised, winglike ridge on T_3, running almost throughout the terminal part. If this latter is opened still more, an almost complete conformity with the features in the Skate will be seen; and thus it will be sufficient to point out the deviations. These deviations are confined to the ventral side, and are chiefly as follows: 1) the membrane covering the inner, dorsal surface of the piece T_3 medially of the cutting edge, has very few and long, obliquely situated, low dermal folds (that may easily be overlooked); 2) the fold *da* is shorter (shortened distally), softer, in the middle of its distal part it projects in a more tongue-shaped manner, upon the whole most like that in *R. radiata*; it is as in this and in *batis* without any inner skeletal support; 3) the two projections corresponding to the skeletal parts *Tv* and *Tv₂*, are somewhat longer, so that they stretch distally over the opening of the ventral recess which thereby gets a somewhat other appearance than in *R. batis*.

Raja circularis Couch.
(Pl. III, fig. 41—44.)

In old males the appendices are said to be somewhat more than ¹/₅ of the total length; in a male of a length of 79.2^{cm}, a breadth of 48,5^{cm} they were 16,5^{cm} long [1].

I have only had the occasion to examine a dried skeleton in the Zoological Museum; this skeleton measures from the snout to the point of the tail 40^{cm}, across the pectorals c. 20^{cm}; the appendix-stem has a length of 6,5^{cm}, the terminal part of 3,7^{cm}, and a breadth of 1^{cm} on the broadest spot. Between the basale and the appendix two pieces are found: b_1 bearing the 8 (7) hindmost rays, and a longer b_2, without rays, as well as a long, plate-shaped *β*, broadest in the fore part.

The ratio between the length of the appendix-stem and $B + b_1 + b_2$ is 3½; the rather narrow, flat, soft terminal part is shorter than the calcified one. The dorsal marginal cartilage reaches forward almost to the beginning of the stem, and ceases behind with a concave, oblique edge, the lateral corner of which is situated much farther forward than the hindmost end of the ventral marginal cartilage, which, as usual, does not reach so far forward. As in *Raja batis* and *clavata*, the dorsal marginal cartilage sends forth a long, thin, pointed, lamellar (calcified) prolongation passing in between the terminal pieces (it is not seen in any of my figures).

The number of terminal pieces is five, besides two dorsal covering pieces. One of these latter, d_1, I suppose to be corresponding to the piece that in *R. radiata* has been marked in the same way; it is long and narrow, spreads distally in a spoonlike manner, and the medial edge of the broader part folds round the appendix-style towards the ventral side, where it is attached to the distal end of the piece T_3 (see fig. 43). The other covering piece, d_3, is firmly connected with the dorsal surface of the terminal piece Td_4 (as is also the corresponding one in *R. radiata*) and is (as in this) distally closely connected with the end of the appendix-style; it is rather thin and flat.

1) Malm, A. V.: Göteborgs och Bohusläns Fauna, 18,,. p. 608.

On the dorsal side are found two real terminal pieces (see fig. 41, 42), Td and Td_i. The former is for a long way connected with the dorsal marginal cartilage and with the appendix-stem; from the hinder end it sends forth a long, round, somewhat finger-shaped, bent process with rounded end, and running obliquely towards the ventral side; on the dorsal side only a little of the basal part of this process is seen, while a larger part may be seen from the ventral side (comp. fig. 44 Td). With the outer hindmost corner of Td a piece Td_i articulates, dividing behind into two branches, an inner one, short and soft, attaching to the appendix-style, and an outer one, hard, compressed, produced to a fine point (corresponding to the free thorn on the corresponding piece in *R. radiata*); this latter branch is best seen from the ventral side (fig. 44, Td_2), as it is dorsally hidden by the covering piece d_3.

The ventral lip shows three pieces: Tv, Tv_2, and T_3.

Tv (fig. 42) is slender, anteriorly connected with the terminal end of the ventral marginal cartilage, which is folded round to the dorsal side; next it follows for a long way the appendix-style, then folds ventrally round this as a rather thin prolongation (fig. 44), and ends finally with two small, diverging points at the opposite edge of the style (comp. *R. batis*). This piece Tv is in its foremost part laterally connected with the very large Tv_2. This latter is somewhat half-moon-shaped, and is attached with its foremost concave edge to the ventral marginal cartilage; it sends forth two processes; a short, truncate one close to the medial edge, and laterally of this a long one, bent in the free end like a hook (comp. *R. batis*), the point of which is turned into the appendix-slit (dorsally); the piece Tv_2, on its inner side, towards the slit, is of a flat, spoonlike shape.

T_3 (fig. 43) is narrow, falcate, and its foremost end is situated under the lateral edge of Tv_2 in the appendix-slit, between Tv_2 and Td; in its hindmost third part it bears on its medial, concave edge a process corresponding to the articular head on T_3 in *R. radiata*, but in the present species it passes into a sharp, winglike edge stretching to the distal end of the piece; the lateral, convex edge of T_3 is sharp and cutting.

Holocephala.

In the males of the Holocephales, as is well known, three particular organs are found that are supposed to subserve the copulation, viz.: 1) the peculiar cephalic organ[1] provided with dermal teeth, 2) the pelvic appendages, i.e. the two organs placed in a ventrally open pouch on each side before the ventrals, and whose skeleton is connected with the pelvis by an articulation; and 3) the ventral appendages. Only the two last-mentioned sets of organs, and especially the ventral appendages, which correspond to those of the Plagiostomes, will be mentioned more thoroughly.

[1] This, however, is wanting in the genus *Harriotta* Goode & Bean, the appendices of which are also said to be small and simple ; of its pelvic appendages nothing is said Oceanic Ichthyology: Mem. Mus. Comp. Zool. Harvard Coll. vol. XXII, 1896. p. 32.

Chimæra monstrosa L.

(Pl. I, fig. 14, 15; pl. VI, fig. 69 71.)

The larger part of the appendix[1]) is free of the fin, and the appearance consequently differs rather much from that of the other Plagiostomes; this free part is almost as long as the ventral fin itself in its largest extent (from the point where the foremost part of the fin arises from the body, to the end of the much produced lateral corner). The appendix may also here be divided in a shaft and a terminal part; the shaft is thick, short, only about half the length of the terminal part; its inner contour is straight, the outer one very convex, whereby the appendix gets some resemblance to the part of the human leg below the knee, with a very prominent calf. On the dorsal side the a p p e n - d i x - s l i t runs throughout the free part of the length of the shaft; anteriorly it begins already at the connection with the ventral side of the body as a little roundish opening, the circumference of which is partly supported by the inner skeleton, and consequently it is only anteriorly a little dilatable; from this opening the slit, bent about in the same manner as the lateral contour of the shaft, runs to the base of the terminal part, where it reaches close to the medial edge, and from here it passes on into the terminal part along this edge. In front, behind the described hole, and posteriorly, where the slit passes into the terminal part, its lips can only with difficulty, or not at all, be opened on account of the stiff inner skeleton, but in the rather long interspace it is easily opened, as the lips are composed of soft parts (muscles); in a specimen before me the two concerning, normally tight spots of the slit are c l o s e d by the coalescing of the skin; in another specimen the case is the same, only to a less extent, with the right appendix. The skin of the shaft is naked, smooth, thin, and slightly pigmented, so that the muscles and their arrangement can be distinguished rather distinctly through it.

The long terminal part is composed of three branches[2]) a medial one (b*) in immediate continuation of the straight medial edge of the stem; a dorsal one (b**), lying quite close to the lateral edge of the foregoing, commonly only separated from it by the very narrow continuation of the appendix-slit; in one single instance, however, I find the skin coalesced for a considerable part of this slit, so that these two pieces only towards the point can be separated; finally a lateral branch (b***), rather free of the other two. These three branches are generally of almost equal length; sometimes the medial one is a trifle longer than the others; they are stiff, and in their whole length supported by skeleton; the medial one is covered with a fine, but firm, thin skin, through which the skeleton is seen very distinctly; it is rounded on its inner, medial surface, and ends in a little, swollen knob; the lateral side is flat, and pressed into a furrow in the skin of the dorsal branch. This latter branch and the lateral one are more or less completely wrapped by a soft, loose, and tooth-covered skin, by which they are

---

[1]) The copulatory appendages have been described, more or less completely, by a rather large number of authors, of whom I shall only mention: Gunnerus: Om Hav-Katten, Det Throndhj. Selsk. Skr., 3. 1763, p 299, pl V VI; Kröyer, l. c. p. 798 seq. Lilljeborg, l. c. p. 518; Duméril, l. c. p. 681; Morean, l. c. p. 463; the descriptions in Gegenbaur l. c. p. 153, and v. Davidoff l. c. p. 453 are very complete. None of these authors mention the glandular bag.

[2]) In *Chim. collici* Benn. living in the Pacific Ocean (at the coast of California), the terminal part is said to have only two branches (Duméril l. c. p. 681, Goode & Bean l. c. p. 32); Bashford Dean (Fishes living and fossil, 1895) gives however, p. 107, fig. 116, a drawing showing three branches, the medial one of which is invested at the point with dermal teeth, and besides by an articulation separated from the other part; this latter fact may perhaps be caused by an accidental damage.

made thicker and, towards the end, enlarged in a clavate manner, when compared with the medial one; the lateral branch in particular is often distally much swollen. The dermal teeth are fine, a little bent thorns, all with the points forward, towards the base of the organ. The lateral branch does not contribute to the bordering of the appendix-slit of the terminal part, this slit running only between the medial and the dorsal branch [1].

In three specimens of the respective length of 78^{cm}, 77^{cm}, and 70^{cm}, the measures were:

The length of the appendix from the fore edge of the cloaca   7,5^{cm}, 10,5^{cm}, 6,5^{cm}
The free part of the shaft . . . .   2,3^{cm}, 2,6^{cm}, 2,3^{cm}
The terminal part . . . .   4,5^{cm}, 6^{cm}, 4,1^{cm}
The breadth (on the broadest part of the shaft) . . . .   1,1^{cm}, 1,6^{cm}, 1,1^{cm}
The breadth (on the middle) of the terminal part . . . .   0,7^{cm}, 1,1^{cm}, 0,7^{cm}

The pelvic copulatory appendage has in all three specimens a length of .   2,1^{cm}
    —   —   —   —   . . —   —   a breadth of   0,6^{cm}.

In one pair of ventrals, kept in spirit, and skeletonized until the terminal part, belonging to a specimen the total length of which I am not able to give, the appendix has had a length of more than 9^{cm}, the terminal part of almost 6^{cm} by a breadth on the middle of 1^{cm}, at the end of 1,5^{cm}; the skeletonized pelvic appendage is 2^{cm} long, and 1^{cm} broad.

The skeleton. The pelvic arch is divided in the middle line, so that it is composed of a right and a left piece; behind, dorsally above the articulation with the ventral, each of these pieces is prolonged to a considerable process; on the foremost convex edge the peculiar, movable, foremost copulatory appendage, the Sägeplatte (Gbr.), is articulated; the skeleton of this appendage is composed of one piece, the medial edge of which bears a row of (5—7) large, crooked, finely pointed dermal teeth; when in rest this piece is turned against the ventral surface of the pelvis which is hollowed like a spoon, and then only the toothless edge laterally of the row of teeth is seen in the opening of the pouch.

The fin-stem consists of a short, flat basale B bearing all the rays (the foremost broad marginal ray (R) is coalesced with it), a b_1, a good-sized β, and the appendix-stem b [2].

b_1 is not much shorter than the basale, with which it is connected in a rather movable joint; on its medial side it is flat and broad, on the lateral side longitudinally concave; dorsally it forms a narrow edge, forward produced into a large process x, which by a lateral incision is made to form the inner bordering of the above mentioned opening, with which the appendix-slit begins; the other part of the dorsal edge of b_1 is somewhat laterally bent, and bears a rather firm margin of connective tissue; the ventral edge is straight and rounded.

The piece β is tolerably triangular, but with curved sides; it is much curved, and situated in

---

[1] In *Chimæra affinis* Cap. the appendices, according to Goode & Bean (l. c. pl. X, fig. 31, 35), are three-branched as in *Ch. monstrosa*, but else they seem to differ rather much from those of this latter. The figures, however, are not distinct enough to get a clear notion of the facts.

[2] In the figures of Gegenbaur l. c. pl. XVI. fig. 22, 23. and of v. Davidoff, l. c. pl. XXIX. fig. 19. pl. XXVIII. fig. 3, 4. these skeletal pieces are marked in such a way that: $b_1 = \begin{cases} b & \text{Gbr.} \\ c' & \text{D.} \end{cases}$, $\beta = \begin{cases} r' & \text{Gbr.} \\ c_1 & \text{D.} \end{cases}$, $b = \begin{cases} b_1 & \text{Gbr.} \\ c_2 & \text{D.} \end{cases}$

such a way, that its concave side like a roof covers the lateral edge of the piece b_1; with its hind-most corner it is by means of tight connective tissue attached to the lateral surface of this piece; its medial edge is free, and forms the limit of the tight foremost part of the appendix-slit, as it also, together with the process x of the piece b_1, contributes to the bordering of the aperture, in which the slit opens anteriorly. We find thus between *Chimæra* and most likely all the Holocephales (*Callo-rhynchus* shows the same relations) on one side and the Plagiostomes on the other the great difference that the appendix-slit anteriorly stretches over the piece β, and on the dorsal side separates this piece from the other parts of the stem skeleton.

The appendix-stem b is joined to b_1 by an only slightly movable articulation, and forms the whole terminal skeleton; no secondary cartilages are found, and consequently the terminal part cannot be directly homologized with that in the Plagiostomes. The part of the appendix-stem lying in the shaft, is short, medially flattened; its medial surface is continued directly in the prolongation forming the medial branch of the terminal part; in the lateral surface is found a furrow-shaped hollow contin-uing the furrow in b_1; both edges of this furrow are elevated and bent towards the concavity, what especially applies to the ventral edge, which rises very much, bends quite over on the dorsal side, folding over the edge of this latter, and lying close to the medial continuation, following this latter quite to the end as the skeleton of the dorsal branch of the terminal part; laterally it forms the car-tilaginous prolongation supporting the lateral branch of the terminal part.

That the cartilage of the medial branch of the terminal part is homologous with that part of the appendix-stem, which in the Plagiostomes I have called the end-style, is an obvious conclusion, and admits of no doubt. At a first glance it seems also obvious that the plate-shaped, folded ventral edge with the two other branches must be corresponding to the ventral marginal cartilage in the Plagiostomes, which latter frequently in Sharks recalls it by the plate that is bent in a similar manner; it might even be tempting to continue, and take the two branches, the dorsal one and the lateral one, to represent two terminal pieces (resp. T_2 and T_3) coalesced with the ventral marginal cartilage; or it might be supposed that this part of the skeleton in *Chimæra* was representing a stage where the terminal pieces had not yet been articulated off as independent parts[1]. But a closer exa-mination shows that the idea of these homologies must be dismissed; the folded ventral edge with its two prolongations is in *Chimæra* absolutely one with the other appendix-stem, consists like this of the same kind of hyaline cartilage, which is corroborated by a transverse section; as a homologon of this structure in *Chimæra* the question can only be of the more or less distinct ventral bordering ridge on the appendix-stem in the Plagiostomes, bearing and continuing the ventral marginal cartilage (see for inst. the Greenland Shark). In the firm, liplike edge of connective tissue, which in *Chimæra* follows the dorsal cartilaginous edge of the appendix-slit, an indication is found that may possibly be regarded as homologous with the dorsal marginal cartilage in the Plagiostomes.

The muscular system. I shall only here describe the muscles that are of importance with

---

[1]) This has also been intimated by Gegenbaur l. c. p. 455: at the same place he intimates that his supposition that the terminal pieces in the Sharks are transformed rays may possibly be wrong, since in *Chimæra* the branches are in contin-uous connection with part of the stem-skeleton.

regard to a comparison with those mentioned in the Plagiostomes as belonging to the appendix; as to the other muscles I may refer to the thorough description by v. Davidoff (l. c. p. 473 seq.).

Between the two halves of the pelvis a broad band (fig. 69—71, s) is stretched, which, as it were, supplements the hinder surface of the pelvis; anteriorly this band is attached along the whole concave posterior edge of the pelvic arch, and laterally it reaches almost to the articulation between the pelvis and the basale; in the median line it is somewhat thickened as a firmer tendinous stripe. From the whole ventral surface of this band as well as from the ventral surface of the pelvis arises the ventral layer[1]) of the group of muscles representing the *M. adductor (et depressor) pinnæ* in the Plagiostomes; in the middle line a stripe broadening somewhat backward, is left uncovered (see fig. 70). This muscular layer is composed of bundles that are distinctly seen distally. Of the medial and hindmost fibres of this layer only the deepest-lying are attached to the ventral side of the basale, to the thickened medial edge of this piece, from which edge the ventral ray-muscles (*Ra*, fig. 70) arise; otherwise the greater portion of the medial fibres of this muscular layer is attached to these ray-muscles until a line of insertion, distinctly seen in fig. 70. The foremost and lateral parts of this muscular layer pass, without any bordering — neither in the depth —, into the ventral ray-muscles, as is also the case in the Plagiostomes[2]). The other muscular mass[3]) which together with the foregoing one forms the *M. adductor* in the Plagiostomes (fig. 69, 71, *A*), arises from the dorsal side of the above mentioned tendinous band, as well as from part of the dorsal surface of the pelvis (viz. until the slight crest that separates it from the muscle *m* of the pelvic appendage); this layer is thicker than the ventral one, and attaches to the thickened medial edge of the basale and to the piece b_1, especially with a powerful portion of fibres to the large process *x* of this latter piece; on the ventral side it reaches to the muscle *D*, which corresponds to the *M. dilatator*, and will be more particularly mentioned hereafter. A special *M. extensor* has not been separated.

A far as I am able to see, only two[4]) muscles are found on the appendix-shaft, one corresponding to the *M. dilatator (D)* in the Plagiostomes, the other to the muscular investment of the glandular bag (inclusive of the outer lip-muscle), *M. compressor sacci (S)*.

The *M. dilatator* arises anteriorly with its ventral portion from the hinder end of the basale, but with its other parts only from the piece b_1, at some distance from the articulation between this piece and the basale. Almost all the fibres run straight from before backward; only on the ventral side some of them bend laterally; they are attached on *b* close to the base of the lateral and medial branches of the terminal part, and a few fibres go to the skin covering the skeleton; on the

---

[1]) Oberflächliche ventrale Schicht, *sst*, fig. 16, 17, pl. XXIX, v. Davidoff.

[2]) I find upon the whole that the difference as to the arrangement of the ventral part of the *M. adductor* in Chimæra and in the Plagiostomes is only in degree; in many of these latter (*Scyllium*, *Pristiurus*, the Rays), the superficial part of the ventral layer of the *M. adductor* stretches quite over the ventral side of the basale and more or less out on the ray-muscles. v. Davidoff describes this ventral muscular layer in *Chimæra* as stretching considerably farther laterally on the fin than is really the case; and his words (l. c. p. 474): "Zum Basale hat er gar keine Beziehung" etc, are not correct.

[3]) The "pelvico-basale Fascrn" of v. Davidoff, fig. 15, 17, *Pb*; they do not, however, as he thinks, arise exclusively from the pelvis.

[4]) v. Davidoff, l. c. p. 480, counts three, which he moreover calls "vollkommen gesondert", viz. a "Flexor", an "Adductor", and an "Abductor"; in three specimens of *Chimæra* that I have examined, I have not been able to find a real separation between the two first-named; but even if such a separation might appear, it will be of only slight importance with regard to a comparison with the Plagiostomes (as surely also with regard to its functions); at all events. "Flexor + Adductor" v. D. is = *M. dilatator*; the "Abductor" of v. Davidoff is the muscle of the glandular bag.

dorsal side (fig. 69) the muscle stretches considerably farther backward than on the ventral side, reaching to the spot, where the appendix-slit passes to the medial side.

The muscle of the glandular bag, *M. compressor*, arises from the lateral edge of the piece β (see fig. 69), and is inserted on the lateral surface of the piece b_1, and on the appendix-stem, as also on the folded ventral edge of this latter. The fibres seen on the ventral side (fig. 71), pass from the edge of β round the calf, running obliquely or transversely, so that part of them is inserted perpendicularly on the appendix-stem; those seen from the dorsal side, on the contrary, run straight from before backward, and they form the lateral limit of the appendix-slit, and are attached where the edge folded from the ventral side, is prolonged as the dorsal terminal branch (fig. 69). The opposite, medial, lip of the appendix-slit is formed by the *M. dilatator*.

Into the described, very voluminous muscle the dermal fold representing the glandular bag in the Plagiostomes, sinks from the dorsal side through the appendix-slit. This structure has here evidently remained in a state of development as that, with which it begins in the Plagiostomes; by a transverse section we see that the bag may in reality be called rudimentary, as it only fills very little in comparison with the powerful wrapping muscular mass. If we imagine this invagination to grow very much forward and ventrally, we may get a structure resembling that in the Plagiostomes; part of the bag will then be situated on the ventral surface of the fin itself, and the muscular coating will, as it were, be extended to a thinner wrapping layer, while the part keeping its position along the outer edge of the slit, will retain its original appearance and become the lip-muscle. This dermal bag, which in *Chimæra* is so small, and whose inner surface is quite smooth and shows no special gland, can nevertheless give plenty of secretion; this fact is proved by the abundance of fluid, partly filling the bag, partly adhering to the branches of the terminal part, and also filling the corners between the base of the fin and the body; on the last-mentioned place it may be supposed to have flown from the foremost, larger, roundish opening of the appendix-slit.

I have not a quite clear understanding of the influence of the muscles of the appendix-shaft on the terminal part; however, I think it likely that by a contemporaneous action of both the said muscles a - probably rather slight - distension of the three terminal branches may be brought about, the *M. dilatator* acting on the medial branch, the *M. compressor* on the two others; by this action the continuation of the appendix-slit between the medial and the dorsal branch would be opened. That also here the *M. compressor* will serve for the pressing out of the secretion of the glandular bag, seems to me to admit of no doubt.

As to the pelvic appendage (fig. 70, 71, *p*), to which nothing corresponding is found in the Plagiostomes, it is in *Chimæra* rather simple; its contour is tolerably spoon-shaped, and it bears on the surface that in the position of rest is turned ventrally (but which will accordingly be turned dorsally, when the organ is directed forward), a soft, loose, unpigmented or slightly pigmented dermal cushion, while the membranous skin of the opposite surface fits tightly to the skeleton. For moving this organ has only one muscle (fig. 69, *m*), by which it can be raised in such a way as to come out of its pouch[1], when it is able to take hold with the toothed edge. This muscle is very powerful;

---

[1] Comp. also Garman: On the Pelvis and External Sexual Organs of Selachians etc, Proc. Boston Soc. Nat. Hist, Vol. 19, 1876—78, p. 199.

as it has no antagonist, the resistance of the surrounding skin, and, I think, also the pressure of the abdominal muscles over the base of the pouch, must be regarded as the cause why the organ folds back and is hidden in the pouch, when the muscle *m* is relaxed. The way in which this muscle is attached, has been described more in detail by v. Davidoff (l. c. p. 479).

Callorhynchus antarcticus Lacép.

The appendices have been briefly mentioned by Duméril (l. c. p. 681) as follows: «Ceux des Callorhynques .... consistent en des prolongements cutanés, enroulés de manière à former une paire de cylindres creux et irréguliers que soutiennent des cartilages flexibles ; the foremost pair of organs, which are enclosed in the pouch, and have here a far more complicated structure than in *Chimæra*, have been more particularly described. The same organs have later been mentioned, though still rather briefly, by T. Jeffery Parker, in a kind of preliminary note[1]), in which is found the rather bold hypothesis, that these anterior appendages in *Callorhynchus* are representing a middle pair of limbs·, they being understood as serially homologous with the real appendices; thus *Callorhynchus* (and the *Chimæra* in general) should (but to be sure only in the males!) show the remains of a hexapod stage·. The real appendices (posterior claspers) are only mentioned with a few words to the effect that they correspond to those in the Plagiostomes, as they occur in the same position, have the form of a plate rolled longitudinally into a tube, and are supported by a prolongation of the basipterygium .

In a specimen (in the museum in Copenhagen) of a length of abt 70ᶜᵐ I find the following measures:

The length of the (real) appendix from the fore edge of the cloaca 8,5ᶜᵐ

The length of the terminal part 5ᶜᵐ

The breadth across the base of the shaft 1,7ᶜᵐ

·   —   —   · middle of the terminal part . . . . 0,8—0,9ᶜᵐ.

As to the habitus the appendix at a first glance reminds more of that in the Sharks than of that in *Chimæra*; but a closer examination shows a very near relation to the latter; it is only the terminal part not being split into branches, that causes the apparent resemblance to the Sharks; the shaft corresponds in shape quite to that in *Chimæra*, and is, as in this latter, covered with a thin, naked skin, through which the extension and form of the muscles may be distinctly discerned; on the terminal part there are, as in *Chimæra*, no muscles at all; but here the skin is everywhere thin, and is lying immediately over the skeleton, so that a reliable view may be got of the structure of this skeleton — unfortunately I could not skeletonize the only male specimen of the museum. The terminal part is somewhat dorso-ventrally flattened with rather parallel sides, only a little tapering towards the rounded end.

On the dorsal side the appendix-slit runs as a narrow slit, beginning, as in *Chimæra*, with a rather large opening at the base, close to the abdomen; this hole is supported by skeleton to the same extent as in *Chimæra*; from here the slit runs in a curve through the shaft into the terminal

[1] Notes from the Otago University Museum, VIII On the Claspers of Callorhynchus. Nature, vol. 33, 1886, p. 635.

part, where it passes over the medial edge on the ventral side, and here it ends in the shape of an S; thus the whole slit is formed like a cork-screw. Immediately behind the hole the edges can only with difficulty be separated on account of the stiffness of the skeleton; but in the terminal part the slit is easily opened on account of the thinness of the skeleton, which is here like a thin, convoluted shaving, which may to a certain degree be unrolled. The inner, tubular hollow of the terminal part, as well as its opening at the point is completely stuffed with secretion, which also fills the hole at the base as well as the nooks between the appendix, the base of the fin, and the body.

As to the skeleton, I think it pretty sure that in the shaft it is as in *Chimæra*; as we find a rather movable joint before the hole, the surroundings of which seem to be quite as in *Chimæra*, we may be justified in supposing the basale to end here; somewhat out on the shaft we find another, little movable joint; accordingly

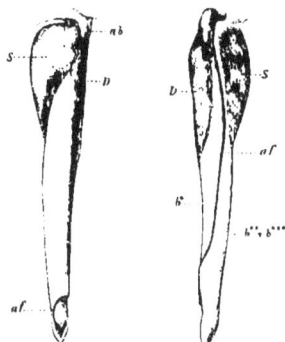

Fig. 27.   Fig. 28.

Fig. 27. *Callorhynchus antarcticus*. The right ventral appendage from the ventral side; a little reduced. *ab* abdominal pore.
Fig. 28. The same from the dorsal side.

the piece b_1 is found between these two points; on the ventral side the distal border of this piece is distinctly marked by the cessation of the inmost part of the muscular mass of the glandular bag (the calf); the other part of the skeleton then must be the appendix-stem; this seems here to be formed like a convoluted leaf, in which no separation into branches is found; such branches, no doubt, would be discernible through the membranous skin, if lines of separation really existed (the only place where such a line of separation might perhaps be found, is along the lateral edge of the ventral side, where a longitudinal furrow in the skin is found in both appendices, but I can find no mobility along it, and take it therefore to be due only to the skin). If we imagine deep incisions in this cartilaginous leaf, the three-branched form in *Chimæra* might arise; on the other hand we may from the three branches in *Chimæra* (see pl. I. fig. 14, 15) easily reach the structure in *Callorhynchus* by imagining a coalescing on the dorsal side (fig. 14) of b^{**} and b^{***}, on the ventral side of b^{***} and b^* (fig. 15).

The muscular system, with regard to the appendix-shaft, is evidently as in *Chimæra*; $M.$ *dilatator* (D) is easily recognised; its chief portion is situated dorsally $(M.$ *abductor* v. David.), and originates on b_1 while the inner and ventral portion $(M.$ *flexor* v. D.) also here arises farther forward on the basale, and does not reach so far backward; further the large muscle (N) of the glandular bag, which in no respects shows other relations than in *Chimæra*, with the only exception that is is a little shortened ventrally.

The foremost copulatory organs, the pelvic appendages, are very remarkably formed, and much larger than in *Chimæra*. The pouch in which they are hidden, is therefore also much larger; the entrance of this pouch forms, when closed, a longitudinal slit (abt. 2,5^{cm} long), and is situated laterally, immediately before the ventral fin. The chief part of the organ enclosed in the pouch, corre-

10*

sponds to the Sägeplatte [1]) of the *Chimæra*; it is here more lengthened, with a longer shaft broadening in the free end to an obliquely cut off, rather thick plate; this plate is on its (when in the position of rest) ventral surface towards the edge invested with numerous, flat, sharp, comb-shaped dermal teeth, of which those nearest the edge are the largest. Each tooth is almost fan-shaped with the edge divided into 5–7 pointed comb-teeth, of which the middle one is the largest (comp. Duméril l. c. pl. 14, fig. 2, 2 a). The teeth continue, somewhat smaller, along the whole (dorso-) medial edge of the Sägeplatte ; they are placed on a dermal lip, which is unsupported by skeleton (and borders the afterwards mentioned canal, into which a sound can be introduced). As far as I can see the teeth continue all the way to the attached base of the organ. The points of all these teeth are turned forward.

On the opposite surface (the dorsal one when in the position of rest) this Sägeplatte is provided with a rather curiously elaborate system of large dermal folds partly supported by an inner cartilaginous skeleton [2]). From the dorsal edge of the organ a large, folded dermal plate projects on either side. 1) The lateral one of these plates somewhat recalls a human ear, and is with its free edge folded towards the dorsal side of the «serrated plate»; the proximal part of this membrane is supported by a particular cartilage, while the distal part which is very much folded, has no inner skeleton. 2) Next another dermal leaf arises from the dorsal and lateral surface of the «serrated plate», opposite to the former; it is folded the other way, and situated between the «serrated plate» and the first leaf; it has no skeleton. 3) The second dermal leaf projecting from the «serrated plate» towards the medial side, is anteriorly grown fast to the inner wall of the pouch; its free edge is folded in such a manner, that it forms a kind of bag; it contains no skeleton, but where it posteriorly is united with the lateral leaf (1) at the dorsal edge of the «serrated plate», these two leaves, by a rolling of their common hindmost part, form a tube supported by a stiff cartilaginous skeleton; the free end of this skeleton projects some way past the end of the «serrated plate» (cp. the figure of Duméril). A sound inserted into this tube, can be brought far into a deep, dorsally open canal along the dorsal edge of the «serrated plate»; on the medial side the bordering of this canal is formed by the above mentioned teeth-covered dermal lip. 4) Finally a big, white, ovate body projecting from the medial wall of the pouch, is found outside the bag formed by the inner dermal leaf (3). This evidently is a glandular body [3]), the opening (or rather openings) of which seem to be inside the bag formed by the inner leaf (3), in the foremost, inner corner of this bag. From this gland proceeds the very abundant secretion filling the space between the «serrated plate» and the described elaborate dermal folds, as also the peculiar tube, evidently the excretory duct for this secretion. As

[1]) As far as I have been able to determine without dissection, this part in the specimen in hand has a length of ca. 3.5cm, a breadth of ca. 2cm across the broad terminal part.

[2]) Only the cartilaginous skeleton has been rendered — and scarcely quite completely — in the mentioned figure by Duméril, as also this skeleton only is mentioned in the text (l. c. p. 682); this work, therefore, gives only a very incomplete idea of the whole organ; the same may also be said of the short communication by Parker in Nature ; upon the whole it is very difficult, without drawings and dissection, to give a tolerably clear survey of these complicated structures.

[3]) Parker evidently has also seen this glandular body: «In connection with the sac is a gland secreting a lubricating fluid, and closely resembling the well-known gland of the Elasmobranch claspers (which gland, however, is not found in all Elasmobranchii). To this is added the interesting observation: «In the female, although the clasper itself is absent, a small glandular sac occurs in the corresponding position». Garman (l. c. p. 200) has, earlier than Parker, seen the gland, and given a very short and incomplete description of the pelvic appendages. He thinks that the above described cartilaginous tube serves for conducting the secretion into the groove of the penis (i. e. the appendix-slit), when it is turned forwards, and through the latter the fluid is conveyed to the oviducts of the female». The supposed turning forward of the appendix,

to the signification of this secretion as well as to the use of the whole organ we can only form rather vague conjectures.

Of the mobility and muscular system of the pelvic appendage Parker gives the following information, showing that the same muscle is found here as in *Chimæra*: The Clasper is exserted by the action of a strong muscle arising from the inner face of the pubic cartilage and passing over its anterior border to be inserted into the principal cartilage (the serrated plate) of the clasper. The plane of movement of the organ is nearly horizontal .

3. Which is the Function of the Appendices Genitales?

That the ventral appendages are peculiar to the males of the Chondropterygians is an old and widely known fact they have already been mentioned by Aristotle[1], and at the present day most fishermen distinguishes the male from the female by means of them[2]. Everybody then agrees that these organs in some way or other subserves the copulation; but till recently we have had no real observation of the copulation of Chondropterygians[3], and we have therefore been obliged to form our opinion of the use of these organs from their structure. Many authors — Rondelet[4], I think, as the first have thought the appendices only to be organs for clasping the female during copulation, and therefore names as Holders , Claspers , Haftorgane , Halteorgane , Klammern , and the like have been generally used; as a consequence of this idea they have always, I think, been considered to act as a kind of prehensile organ, which might cling to some part of the body of the female outside, and thus hold it fast[5]. Others, on the contrary, have supposed that these organs have to be introduced into the sexual organs of the female; but their action there has been interpreted in several ways. Almost all the earlier authors, as Linné, Artedi, Willughby (Ray), Klein, Battarra, Gunnerus have thought that they convey the sperm, and called them *Penes, Mentulæ*, or *Membra genitalia*, and with regard to their existing in pairs, some of those authors refer to the Snakes, which analogon also seems to be rather obvious. After the appearance of the works by Bloch, the first-mentioned idea of the appendages as mere external clasping organs gained many

however, cannot take place, and so the whole supposition has to be dropped. Garman does not mention the structure of the appendix itself.

[1] *Ἱστορία περὶ ζῶων.* Ed. by Aubert and Wimmer. Leipzic 1868, vol. I. p. 455, Chap. V, 5, § 15.

[2] Lorenzini (Osservazione intorno alle Torpedine, 1678), who, it would seem, has only known the appendages in the Rays, declares that they may be found in both sexes. He says nothing of their function. This misconception that they are also to be found in the females — recurs oftener. I think the assertion by A. Fritsch, that in the Xenacanths the old females are also provided with appendages, to be not better founded. (In Bashford Dean: Fishes living and fossil (Columbia Univ. Biol. Series, III) 1895, a figure is found on p. 73, representing General Anatomy of shark (?) , and this ♀ is provided with claspers !).

[3] The lively description by La Cépède (Histoire naturelle des Poissons, T. I. p. 251 55) of a copulation between two Sharks, is evidently not founded on observation. His description and construction of the appendices (l. c. p. CXLIII, p. 70, p. 273) are based on the essays of Bloch.

[4] Libri de piscibus marinis, 1554, Lib. IIII, p. 89: Mares cartilaginei fere omnes circa podice appendices duas habent quibus coire creduntur. At ego illas sæpe multumque contemplatus, non video quo pacto his coire illi possint; potius igitur ad retinendas foeminas factas esse arbitror .

[5] This is decidedly said by Bloch (Schr. Berl. Ges. vol. 6, pp. 379, 384), whose excellent representation seems to have influenced very many authors; further by Home (Phil. Tr. 1809, p. 207, and 1810, p. 206), by Cuvier & Valenciennes (Hist. nat. des Poissons I. I p. 376) by Treviranus (Tiedemann u. Treviranus Zeitschr. für Physiologie vol. 2, 1828, p. 9 (in the explanation of the figures]), by Duméril (l. c. p. 241), and others.

adherents, but the notion that they are real penes, i. e. organs conveying the semen, seems never to have been fully superseded by it; Blainville advocates this opinion[1], and later we find it in Mayer[2], in Leydig[3], Stannius[4], Steenstrup[5], L. Agassiz[6], Davy[7], and (partly) Günther[8]); this opinion, after all, is perhaps to this day the most widely spread; it is also rather obvious, and analogies from other groups of animals present themselves, as it were, spontaneously[9]). This interpretation of the appendages as the direct conveyers of the semen, however, meets with difficulties, which partly have been correctly seen by several authors; some of those then have adhered to the opinion that they are introduced into the cloaca of the female, but only to be more indirectly subserving the copulation. Thus Geoffroy St. Hilaire[10] characterizes them as clitores, and Petri[11] thinks their chief

[1] l. c. p. 126. Blainville promises a treatise on la structure et les usages de ces appendices dans les raies et les squales , in which he even thinks to have found a connection with the sexual organs proper, what he had not been able to do in le Squale pélérin .

[2] Über die Bedeutung der fussförmigen Anhänge bei Rochen und Hayen, und ihr Wiedervorkommen bei niederen Thieren. Prorieps Notizen aus dem Geb. der Natur- und Heilkunde. vol. 40. 1834. p. 273. Mayer supposes that these limbs by the *Musculi adductores* are brought to the cloaca, receive the semen into the appendix-slit, and convey it on to the terminal part, the opened leaves of which wie ein Blumenkelch embrace the cloaca of the female; further he imagines that the copulating animals wahrscheinlich von einander abgewendet sich befinden (Petri l. c. p. 291 renders the description by M., but in more respects incorrectly).

[3] l. c. p. 86. Die sogenannten Haftorgane erinnern in ihrer gewundenen, rinnenförmigen Gestalt sehr an die äusseren Begattungsorgane mancher Krebse und ich glaube, dass sie ebenso wie diese zum Überpflanzen des Samens nach den weiblichen Geschlechtstheilen dienen, wobei dann das Sekret der oben beschriebenen Drüse eine vielleicht die Samenmasse einhüllende oder schützende Rolle spielt .

[4] Handbuch der Anatomie der Wirbelthiere, 2 Aufl., 1854, 1, p. 278, note 5.

[5] Hectocotyldannelsen hos Octopodslægterne Argonauta og Tremoctopus. Kgl. D. Vid. Selsk. Skrifter, 1856, p. 26. «I think, however, ... the analogon to be as obvious, which is found in so many males among the decapod Crustacea, in which a pair of the abdominal limbs are formed as more or less complete tubes, or the analogon, seen in the male Rays and Sharks, where the ventrals, that is to say, active organs of motion, have one side transformed into large ducts of the semen.

[6] 1) Proceedings of the Boston Society of Nat. Hist. Vol. VI, 1856-59, p. 377. 2) Ibid. Vol. XIV, 1871, p. 339. In the first-mentioned place is only found a report of some observations by Agassiz occasioned by a lecture on the egg-development in Rays; he thinks the claspers of the Rays to be real copulatory organs, supposing them to be turned forward and upward, by which turning an opening in them (the larger basal opening of the appendix-slit?) is brought up to the spermatic ducts; it is supposed that they may easily be introduced into the oviduct even to the shell-gland. In the later communication (2) this is more particularly worked out: One ray of each posterior fin is capable of erection and rotation, and is covered with erectile tissue, far too delicate to allow it to be used as a clasper around a body covered with sharp rough spines. In the act these two organs are rotated inward and forward, bringing the furrows on their inner surface into parallel contact, and in apposition with the testes. Being then introduced into the body of the female, their extremities diverge in the two oviducts, and the *glans* being uncovered exposes a sharp cutting instrument, which would injure the organs of the female if she resisted; the male has her, therefore, in complete subjection, and has been observed to strike and wound her with this spine. What was formerly supposed to be the penis is too small, and of insufficient length to accomplish fecundation (viz. the urogenital papilla). The penis consists of the two long flexible finger-like fins, furnished with two projectile spinous appendages as in vipers. (In *Chimæra* the surfaces of the organs are also spinous, as in snakes). The two spines found in cartilaginous fishes are homologous with the *os penis* of mammals. In men this bony part has disappeared, and we have only the soft spongy portions of the organ remaining; the quivering of the legs during connection seems the echo, as it were, of the sensitiveness of the flexible posterior limbs of the skates(!). As the thought of a comparison with the Snakes cannot be said to have been exactly new at that time, so it is also the case with the homology with the *os penis* ; it is already found in Ray (Willughby: De Hist. Pisc. etc. 1686, p. 77). Garman, l. c. p. 199–200, subscribes the opinion of Agassiz .

[7] Already l. c. 1830, p. 149; more decidedly in: Fragmentary Notes on the Generative Organs of some Cartilaginous Fishes (Trans. Roy. Soc. Edinb, 1861; vol. 22, p. 500).

[8] Introduction etc. p. 167. Günther also supposes that the two appendices by being put together may form one canal; he thinks it to be possible that the appendix-slit leads as well the secretion of the glandular bag as the sperm.

[9] Besides to the palps of the Armeina, the thought will easily be led to the limbs that in the Crustacea, especially the Decapoda, have been developed for serving the copulation; not only Leydig and Steenstrup, as has been shown by the above quotations, but also Mayer have thought of these; several other analogies indicated by Mayer are rather distant (even if they be not all so distant as those, attributed to M. by Petri: the thumb-swellings in the frogs, the spur of the Ornithorhynchus — which analogies I have not at all been able to find mentioned in M.).

[10] According to Petri; I have not been able to find the essay in question.

[11] l. c. p. 330. The secondary function, which Petri (in accordance with Bloch) ascribes to them: to serve as an organ of motion making the males more mobile than the females - especially in the Rays - may surely, to say the least of it, be characterized as problematic.

employment to be to act as a kind of dilators of the sexual organs of the female; he imagines them to be introduced until the mouth of the oviduct, whereupon the *M. dilatator* dilates the terminal part, so that the bore of the oviduct is enlarged, and the male also is enabled to draw the female nearer to itself, in such a manner that he can with his urogenital papilla reach into the cloaca of the female, and there discharge the sperm, which from there more easily may penetrate into the mouths of the oviducts that have been dilated by the appendices.

None of the mentioned authors have been able to found their opinions on any observation of the copulation [1]. Only of late we have one, as it would seem, reliable observation, communicated by Bolau [2], by which at all events it may be regarded as an established fact that the appendix is really introduced into the genitals of the female. This observation applies to *Scyllium stellare (catulus)*, and is made in the aquarium of the zoological garden in Hamburgh. Before the copulation the male for about a day kept near the female, and pursued her, but it was not observed in what manner he seized her. During the copulation the female is encircled by the male, the latter, as it were, twisting round her cross-wise; only one appendix, it would seem, is introduced at each copulation, and this appendix, judging by the very incomplete sketch given by Bolau, (l. c. p. 322, fig. 2) must also after the act be somewhat dilated. The copulation itself lasted in two observed cases 20 minutes. Bolau follows Petri with regard to the interpretation of the part played by the appendix on this occasion; but he adds that he is not able to decide, whether the appendix-slit [3] plays a part by the conveying of the semen.

This observation, as far as I know, stands hitherto quite alone; it seems to me to be of no small interest, although it decides nothing with regard to the most important question, whether the appendix really conveys the semen or not. As to this question we are still reduced to draw our inferences from the structure of the organ. This structure seems to me to show with complete certainty that at all events the appendix-slit cannot be the duct of the semen; it is situated in such a way, that it is impossible to understand how the sperm should get into it and follow it, as it, as we have seen, is situated dorsally and laterally, sometimes (for inst. in the Skate) quite laterally; the ventrals are not able to perform a movement of such a nature as to make the foremost opening

[1] Davy and Agassiz, however, — as also several of the earlier authors (for inst. Koudeleti — have known the following remark in Aristotle, which might be indicative of some observations really having been made in antiquity: εἰσὶ δὲ τινες οἱ ἐωραχίναι φασὶ καὶ συνεχρημενα τῶν σκλαρῶν ἔνα ὅπισθεν ὥσπερ τοὺς κύνας (l. c. 5, chap. 5, § 14). In his last communication (1861, p. 500) where Davy rather decidedly declares in favour of construing the appendix as a penis, he mentions some circumstances supporting the notion of an intromission, derived from *Centrina*, as for inst. that the cloaca of the female is large enough to receive the appendix, that it appeared slightly lacerated at its superior commissure, and that the mouths of the uteri protruded, and were red and blood-filled. Garman (On the Skates (*Raja*) of the Eastern Coast of the United States. Proc. Boston Soc. Nat. Hist. Vol. XVII, 1874, p. 170), who, as mentioned, subscribes the opinion of Agassiz, to whom he attributes the credit of the discovery of the functions of the claspers, has observed a fact that adds a little emphasis to his (A.'s) discovery viz: that in virginal Sharks the hindmost end of the oviduct is closed as by a kind of hymen (comp. also Semper: Das Urogenitalsyst. der Plagiost. Arb. Zool-Zoot. Inst. Würzb. vol. 2, 1875, p. 279), or provided with a very small pore; this pore is round in the species of which the male has tapering claspers, and forms a short, horizontal slit in those where the claspers are flat with rounded ends; in the species where the appendix has sharp edges and hooks, the hindmost part of the oviduct and the cloaca is very thick and leathery. In virginal *Mustelus* the oviducts were furthermore found stretching along the dorsal side of the cloaca to a point at the middle of the anus; in grown, impregnated specimens they are open, as if an inch or more had been cut off of the end, and the rectum opens in the cloaca between their openings and the outer one.

[2] Über die Paarung und Fortpflanzung der *Scyllium*-Arten. Zeitschr. f. wiss. Zool. vol. 35. 1881, p. 321.

[3] He wrongly places the appendix-slit on the inner side of the organ and its partly closed state in *Scyllium* seems unknown to him.

of the slit approach the cloaca of the animal itself (as supposed by Agassiz [and Garman]); neither can a turning round the longitudinal axis be effected (least of all a turning of 180 , as would be required in the Skate), and thus any thought of a putting together of the slits of the two sides to form a tube (Agassiz, Günther) has to be dismissed (quite apart from the fact that in some forms — *Scyllium, Pristiurus* the appendix-slit is closed for a long way by coalescing). A putting together of the medial sides of the two appendages may however easily be effected by the *Musc. adductores*, but by this no convenient way for the sperm would be formed; and the observation of Bolau shows moreover that in *Scyllium* only one appendix is used at a time; for the present it may, however, be disputed, whether this is a universal law in all cases and in all other Selachians. Thus it seems that the tubular, or rather semi-tubular form of the appendix cannot directly have anything to do with the transferring of the semen; the most immediate purpose of this form evidently is the transportation of the gland-secretion.

On the other hand the structure of the appendix shows with still greater certainty — quite apart from the observation by Bolau — that the appendix cannot be used for externally clasping the female. For a great part, I think, it is the hooks, claws, or thorns, so often projecting through the skin of the terminal part that have caused or supported this supposition. But an attentive observation of the position and way of moving of these firm parts, as also of the whole constitution of the terminal part, might, as it seems to me, rather easily have persuaded the many adherents of the theory of these organs as claspers , or Klammerorgane , that they are only ill adapted for such a purpose. The skin of the whole terminal part is, as we have seen, often quite naked and soft (the point itself is always so), and the appendix would therefore - as has been correctly pointed out by Agassiz — be badly off with regard to the rough surfaces, with which in most cases it would have to do, and against which it would only be slightly protected by the secretion (Bloch; this secretion would rather be a hindrance for the clasping, as is also remarked by Davy). In the *Raja*-species the hard skeletal parts whose business would be to hold fast the female, only appear within the dilated terminal part, and are wrapped by a specially vulnerable skin, very much like a mucous membrane; consequently, if these parts were to hook on — for which their special shape is in no way adapted — for inst. to the thorny tail of the female Ray (Cuvier & Valenciennes, Duméril), their most immediate surroundings would be much exposed to injury; and if we choose to regard such appendages as those in *Acanthias, Somniosus*, or above all *Spinax*, which, by the hooks, thorns, or claws projecting freely through the outer skin, may for a superficial examination convey the impression of being plain prehensile organs (the dilated terminal part of *Spinax* reminds not a little of a bird's foot!), then any closer examining will show that they cannot be such: the position of these claws is always so, that they cannot catch an object, or clutch it. Besides their movement inward, against each other, when the terminal part is closed, always takes place with small force, by elastic reaction of the connecting soft parts, only to a small degree (and not in all cases) somewhat assisted by muscular action. The erection of these parts on the contrary, when the terminal part is opened by means of the always powerful *M. dilatator*, can take place with great force, and they may with force be kept spread out. I think therefore that there can be no doubt, but that Davy has had an eye for the correct fact (although the Rays especially examined by him, do not present the fact so clearly by far, as do *Acanthias* or *Spinax*), when he supposes the appendices to be organs for intromission and retention like the Penis of the Dog-; only in a hollow these spurs, thorns etc., can be of importance as retentive organs; it is quite evident that they are barbs that are kept stiff, as long as the dilation of the terminal part lasts. Viewed in this way the dermal teeth on the terminal part in *Scyllium*, *Pristiurus* (and *Chimæra*) will also get importance, they being placed with their points towards the base of the fin and raised by the dilation; they will also — although to a less degree — act as barbs. When the dilation ceases, all these barbs – large and small – are laid, and thus they present as small resistance as possible by the extraction from, as well as by the introduction into a hollow. That the object is that they may be introduced and extracted without resistance, is very finely shown in some instances; this, above all, applies to the hook *(Td)* in *Acanthias*; in the position of rest it fits so elegantly into the spoonlike ventral terminal piece *(7?)* as to remind of a surgical instrument[1]. All the appendices are moreover adapted for being thickly smeared with the viscid secretion of the glandular bag, and accordingly being made smooth, by which an introduction into a relatively narrow hollow may be highly facilitated.

I think then that the structure of the appendix shows quite indisputably: 1) that this organ is intended for being introduced into a hollow, and 2) that it is able to fix itself in this hollow by the dilation of the terminal part. In this way — but only in this way — the appendix becomes an organ of retention during copulation. It would a priori be the only reasonable supposition, that the hollow of which the question here can be, must be the genitals of the female; by the observation of Bolau this supposition has been made a certainty, and this gives to his observation its special importance. My opinion then is, that at all events it may be put down as certain that the ventral appendages during copulation serve as retentive organs in the genitals of the female[2]. But this can scarcely be their only function. My opinion is that they must have several functions, among others to awaken the sensuality, and furthermore to open (or at all events to widen) the mouths of the oviducts in virginal females, and thus secure impregnation and facilitate the parturition; and though I cannot imagine that the appendix-slit should form a duct for the sperm, I still think it probable that the appendages in some way or other subserve the conveying of the semen, so that it is not conveyed by means of the urogenital papilla of the male alone. And I also suppose that the secretion of the glandular bag subserves this object. As we have seen, the secretion is in all appendices not only evacuated through the hind end of the organ, in the terminal part, but also in all instances through the opening at the base of the organ, and thus not only the genitals of the female and the appendix itself, but most likely the whole immediate surrounding of the cloaca in both the copulating animals will be lubricated by the secretion. The consequence of this will be that the sperm will easily be mixed with the secretion, and it may readily be supposed that this mixing may have a stimulating influence on the spermatozoids, or act as gathering and conveying

---

[1] Gegenbaur, who does not at all mention the function of the appendix, says of these parts in *Acanthias* (l. c. p. 452): Das Verhalten beider Stücke ähnelt den verdeckten Haken, wie sie als chirurgische Instrumente gebraucht werden .

[2] The old, before quoted observation in Aristotle gains by this view very much in trustworthiness: There are those who assert that they have observed that some of the Selachians hang together behind like the Dogs ; and it lies near to suppose that it is this kind of hanging together , that is suggested by Pennant (Brit. Zoology. New. Ed. 1812. Vol. III. p. 113) of the skate: several of the males pursuing one female- and adhere so fast during coition, that the fishermen frequently draw up both together, though only one has taken the bait .

the semen, preventing it from flowing off in the water. Then the part played by the secretion, would not be restricted to facilitating the introduction of the appendix -- which part I regard as quite incontestable —, and to protect the different parts partaking in the copulation (eventually also the outer skin) against a severe friction; but the secretion would also be of direct importance for the impregnation by yielding a means, as it were, of keeping together the semen and leading it along the appendix into the oviduct.

It must be possible to some degree to test this supposition by examining the way in which the spermatozoids act in relation to the fresh secretion; but unfortunately I have had no opportunity for that[1]. For the present I must leave the value of this and my other suppositions to the testing of others, and own that I have only been able to advance the understanding of the functions of the ventral appendages very little; most of the questions raised by the different, rather complicated structures, especially in the terminal part, must still be left quite unanswered, as also such facts as the large extent of the glandular bag in most Sharks must still appear mysterious[2]. With regard to some of these questions it may be dubious, whether they ever will be solved; but with regard to others, especially the question of the appendages as means of the conveying of the semen, it would seem that they might be solved by observations. It is to be hoped that the future will bring such observations.

Addenda.

I have been unwilling in this translation to make any essential alterations of the original Danish text. This latter was ready printed in August 1898. I regret to say that shortly after I saw that I had quite overlooked a short, but rather essential contribution by A. Schneider to the question of the function of these organs; it is only little more than half a page, and is printed in Zool. Beiträge vol. I, 1885, p. 613). In this contribution he says of the glandular bag: Dieser Sack hat jedoch noch eine andere bisher ganz übersehene Function. Er ist ein Receptaculum seminis, Ich habe bei *Spinax Acanthias* Samen darin gefunden. Die Begattung dürfte deshalb bei den *Plagiostomen* in der Weise stattfinden, dass zuerst das Receptaculum seminis mit Samen gefüllt wird und von da aus mit Hülfe des in den Uterus eingeführten Pterygopodium die Immissio seminis stattfindet. Bei

[1] Hitherto only very little is known of the chemical relations of this secretion. Davy (l. c. 1839, p. 145) says it is neither acid nor alkaline, and that it has a very indistinctly acrid after taste. Moreau, on the contrary, declares it to be acid (l. c. p. 288); this, however, can scarcely be correct, as in this case it would have a bad influence on the spermatozoids with which it will scarcely avoid to come into contact.

[2] For those, who are of opinion that Agassiz has solved the question of the function of the appendages correctly, these bags, perhaps, will not appear quite so mysterious; Garman, for inst. says (Proc. Bost. Soc. 1874, p. 173): That the cavity upon the ventrals, containing the muscular gland, fills so readily with the sperm when the claspers are erected, and that its contents are expelled, upon contraction of the muscles around it, with such certainty to their ends, when restored to their normal position, are evidences that it acts as a forcing or squirting apparatus . I must, however, object against this 1) that I cannot see that the sperm upon the whole can be filled into the bag, still less, that it can be done easily; and 2) that spermatozoids never have been found in the glandular bag, although its contents have several times been subjected to microscopical examination, also with the object of seeking spermatozoids in them.

[3] As it is reported in Biol. Centralbl. vol. III, 1883, no. 7, p. 224, this contribution to the .Beiträge. must have appeared two years before the completing of the said volume.

den *Holocephali*, *Callorhynchus* und *Chimæra* besitzt das Männchen vor dem Pterygopodium jederseits einen sehr verwickelt gebauten Apparat. Derselbe besteht aus einer Tasche, in welcher mehrere, Knorpel enthaltende, mit Widerhaken versehene Stücke hervorgestreckt werden können. Ich fand diese Tasche bei *Callorhynchus* mit Samen gefüllt. Auch bei dieser Gruppe der Elasmobranchier wird demnach der Samen vor der Begattung nach aussen gebracht. Wie freilich hier die Begattung stattfinden wird, lässt sich vorläufig nicht angeben. The essential thing is that Schneider declares to have found sperm in the bag in *Acanthias* and in the pouch of the pelvic appendages in *Callorhynchus*; certainly no proof is given, but we shall have to suppose that Schneider has really found the spermatozoids. Whether these have been numerous, that is to say, whether the bags in question really can be said to have been filled with the semen, of this we know nothing with certainty, and we can — in my opinion — not yet in any way put it down as an indubitable fact that the glandular bag of the Plagiostomes is a reservoir that has to be filled with the semen and by the copulation to eject it. Nothing is said of the way, in which the filling of the bags in question should take place.

I have unfortunately not been able to get a paper by Haswell (Notes on the claspers of Heptanchus. Proc. Linn. Soc. N. South Wales. vol. 9, P. 2, p. 381).

During the time between the appearing of the present essay in Danish and this translation I have received a paper by H. C. Redeke (Onderzoekingen betreffende het Urogenitaalsystem der Selachiers en Holocephalen. Acad. Proefschrift etc. Helder 1898) in which (p. 77) after a representation of what till then was known regarding the appendages and their function, the author declares that he has himself found numerous spermatozoa in the mixipterygoid bag [1] in one single specimen among many examined specimens of *Mustelus vulgaris*. He calls, however, attention to the fact that the bag was not filled, which fact he explains by supposing, either that the animal during its agony might have emptied the bag, or rather that these animals will copulate, as soon as the bag is filled. An observation by another observer, respecting a male *Raja clavata* that had ejected an abundance of semen through the dilated appendices, can scarcely be regarded to be of any value, as there is no proof to the effect that the ejected fluid in reality was semen and not the secretion from the gland. Finally is quoted an observation by Professor M. Weber, which observation the author thinks may be used to explain, in what manner the filling of the glandular bag might be brought about. I shall give the proper words of the author, and else abstain from advancing my strong doubt of the fact:

Deze (Prof. Weber) nam waar, hoe een groote Rog (*Raja clavata*) rondzwemmende in een der bassins, plotseling een groote wolk, vermoedelijk sperma, loosde en vervolgens, misschien reflectorisch, heftig met zijn mixipterygien begon te zwaaien, die daarbij een pompende beweging scheenen uit te voeren. Het is niet onmogelijk, dat ook in de natuur, al is de omweg een allerzonderlingste, het sperma eerst in een groote hoeveelheid geloosd en gelijktijdig door de mixipterygien in den zak opgezogen wordt.

[1] The appellation of Mixipterygium, which has of late often been used in stead of the objectionable Pterygopodium of Petri, is due to Gegenbaur (Das Flossenskelet der Crossopterygier etc. Morph. Jahrb. vol. 22, 1895. p. 146, note).

EXPLANATION OF THE FIGURES.

A: *Musculus adductor.*
af: The appendix-slit.
B: The basale metapterygii.
b: The appendix-stem.
b_1, b_2, b_3, b_4: The stem-joints between the appendix-stem and the basale.
β: The dorsal stem-piece.
D: *Musculus dilatator.*
d, d_1, d_2, d_3: Dorsal covering pieces.
da_1, da: Terminal pieces belonging to the ventral side (in some *Raja*-species).
E: *Musculus extensor.*
g: The end-style, the uncalcified end of the appendix-stem.
h: Horny filaments.
O, O': Fin-muscles arising from the body.
P: The pelvis.
R: Marginal ray.
r: Rays.
Ra: Ray-muscles.
Rd: The dorsal marginal cartilage.
Rd': Process from the dorsal marginal cartilage (in *Raja*-species).
Rd_2: A special terminal piece, added to the dorsal marginal cartilage.
Rv: The ventral marginal cartilage.
S: *Musculus compressor.*
s: A ligamentous septum, serving for attaching part of the *Musc. adductor.*
Td, Td_2: Dorsal terminal pieces.
Tv, Tv_2, T_3: Ventral terminal pieces.
v, v': Ventral covering pieces.

Plate I.

Fig. 1—9. *Somniosus microcephalus.*

Fig. 1: The skeleton of the right ventral, viewed from the dorsal side; considerably reduced.
— 2: The chief piece of the right appendage, viewed from the dorsal side; reduced. *l* the lateral surface; *x* articular surface for attaching the piece *β*.
— 3: The same skeletal part, from the ventral side.
— 4: The dorsal terminal piece, *Td*, from the dorsal side.
— 5: The same piece, from the ventral side.
— 6: The ventral terminal piece, *Tv*, from the dorsal side.
— 7: The same piece from the ventral side.
— 8: The thorn or spur, *T₃*, from the dorsal side.
— 9: The same, from the ventral side.

Fig. 10—11. *Acanthias vulgaris.*

Fig. 10: The skeleton of the right appendage, from the dorsal side; natural size.
— 11: The same skeletal part, from the ventral side.

Fig. 12—13. *Spinax niger.*

Fig. 12: The skeleton of the right appendage, from the dorsal side; a little enlarged.
— 13: The same, from the ventral side.

Fig. 14—15. *Chimæra monstrosa.*

Fig. 14: The skeleton of the right ventral, from the dorsal side; natural size. *x* process on the piece *b₁*; *b**, *b***, *b**** the medial, dorsal, and lateral branches af the appendix-stem.
— 15: The same skeletal parts, from the ventral side.

Plate II.

All the figures represent the skeleton of the appendage of the right ventral fin (or parts of it).

Fig. 16—17. *Scyllium canicula.*

Fig. 16: The skeleton of the appendage, from the dorsal side; natural size.
— 17: The same, from the ventral side.

Fig. 18—19. *Scyllium stellare.*

Fig. 18: The appendage, from the dorsal side; natural size.
— 19: The same, from the ventral side.

Fig. 20—21.   *Pristiurus melanostomus.*

Fig. 20: The appendage, from the dorsal side; natural size.
  21: The same, from the ventral side.

Fig. 22—23.   *Lamna cornubica.*

Fig. 22: The appendage, from the dorsal side; much reduced.
  - 23: The same, from the ventral side.

Fig. 24—27.   *Rhina squatina.*

Fig. 24: The appendage, from the ventral side; reduced.
  — 25: The distal end of the same, from the dorsal side; * indicates the place where the terminal piece T_3 ends, hidden in the ventral marginal cartilage Rv.
  ·· 26: The same part, from the ventral side; the covering piece v removed.
  — 27: The terminal pieces Tv and T_3 figured separately.

Plate III.

All the figures represent the skeleton of the appendage of the right ventral fin (or parts of this skeleton).

Fig. 28—31.   *Torpedo marmorata.*

Fig. 28: The distal end of the appendix-skeleton, from the dorsal side; about natural size.
  29: The same; the covering piece v removed.
  30: The same, from the ventral side.
  — 31: The same, from the ventral side; the covering piece v removed.

Fig. 32—34:   *Narcine sp.*

Fig. 32: The appendage, from the dorsal side; somewhat enlarged; the covering piece v removed.
  — 33: The same, from the ventral side.
  — 34: The covering pieces v and v', from the dorsal side.

Fig. 35—37.   *Rhinobatus columnae.*

Fig. 35: The appendage etc., from the dorsal side; about natural size; the covering piece v removed.
  — 36: The same, from the ventral side.
  37: The terminal point of the appendage, with the covering piece v, from the ventral side.

Fig. 38—40.   *Trygon violacea.*

Fig. 38: The appendage etc., from the dorsal side; about natural size; the covering pieces v and v' removed.
  ·· 39: The terminal part of the same, from the ventral side, with the covering pieces v and v'.
  40: The appendage from the ventral side; the covering pieces removed.

Fig. 41—44.   *Raja circularis.*

Fig. 41: The terminal part of the appendix-skeleton, from the dorsal side; natural size.
  42: The appendage from the dorsal side; the covering piece d_3 and the terminal piece T_3 removed.
  — 43: The terminal part of the same, from the ventral side.
  — 44: The appendage, from the ventral side; the covering piece and the terminal piece T_3 removed.

Plate IV.

All the figures represent the skeleton of the appendage of the left ventral fin (or parts of this skeleton).

Fig. 45—48. *Raja batis.*

Fig. 45: The terminal part of the skeleton of the appendage, from the dorsal side; considerably reduced; the covering piece *d* and the terminal piece T_1 removed; *x* a calcified part of the end-style *g*.

— 46: The same, from the ventral side; all the pieces present.

— 47: The same, from the ventral side; the covering piece and the terminal piece T_1 removed.

— 48: The dorsal covering piece, *d*, seen from the dorsal side.

Fig. 49—52. *Raja clavata.*

Fig. 49: The appendage with all its pieces, viewed from the dorsal side; considerably reduced.

— 50: The same, from the dorsal side; the covering piece *d* removed.

— 51: The same, from the ventral side, with all the pieces present.

52: The same, from the ventral side; the covering piece *d* and the terminal piece T_1 removed.

Fig. 53—57. *Raja radiata.*

Fig. 53: The skeleton of the appendage with all its parts, from the dorsal side; reduced; *x* thickened and calcified part of the end-style *g*.

— 54: The same, from the dorsal side; the covering pieces d_1—d_3 removed.

— 55: The same, from the ventral side; all parts present.

— 56: Part of the dorsal wall of the appendix-slit, viewed from the ventral side; the ventral marginal cartilage and all the terminal pieces of the ventral side, as well as the covering pieces removed. *Rd'* is here an independent piece.

— 57: The terminal part of the ventral marginal cartilage with the terminal pieces Tv and Tv_1, separated from the other skeletal parts, and viewed from the dorsal side (i. e. part of the internal side of the ventral wall of the appendix-slit).

Plate V.

All the figures represent the right ventral fin.

Fig. 58—62. *Somniosus microcephalus.*

Fig. 58: Ventral fin, viewed from the ventral side, of a young specimen, 2^m 50^{rm} long; considerably reduced.

— 59: Part of the same ventral fin, viewed from the dorsal side and a little turned.

— 60: Part of the ventral of a large specimen, seen from the ventral side; considerably reduced. The terminal parts, with the exception of part of the spur T_1, covered by aponeurosis.

— 61: The same, from the dorsal side; part of the dorsal ray-muscles, *Ra*, removed, as well as part of the muscular portion *O* arising from the body; *a* aponeurosis of the *Musc. extensor E*.

— 62: Part of the same, showing the muscles of the appendix, after removing the *Musc. extensor E*, the muscular portions *O* and *O'* (comp. fig. 59), as also part of the glandular bag (comp. fig. 61).

Fig. 63 64.   *Acanthias vulgaris* ♀.

Fig. 63:   Ventral fin from the ventral side; natural size.
  - 64:   The same from the dorsal side; most of the muscles arising from the body removed.

Plate VI.

The figures, except fig. 67 - 68, represent the right ventral fin.

Fig. 65—66.   *Scyllium stellare.*

Fig. 65:   The ventral fin from the ventral side; somewhat reduced; the greater part of the glandular
bag *S* removed; a_1, a_2 special muscles of the appendix; *f* the winglike process.
  - 66:   The same, from the dorsal side; *af* the basal opening of the appendix-slit.

Fig. 67 68.   *Raja clavata.*

Fig. 67:   The left ventral fin, from the dorsal side; considerably reduced; the terminal parts covered
by the aponeurosis.
  — 68:   The same, from the ventral side.

Fig. 69--71.   *Chimaera monstrosa.*

Fig. 69:   The right ventral fin, from the ventral side; a little reduced; the skin on the branches
of the terminal part not removed; b^* the medial terminal branch, b^{**} the dorsal one, b^{***}
the lateral one; *p* the serrated plates covered with its skin.
  -- 70:   The same, from the dorsal side; *m* the muscle of the serrated plate ; *x* process on the
piece b_1.
  71:   Part of the same, from the ventral side; the ventral portion of the *Musc. adductor, A* in
fig. 70, removed.

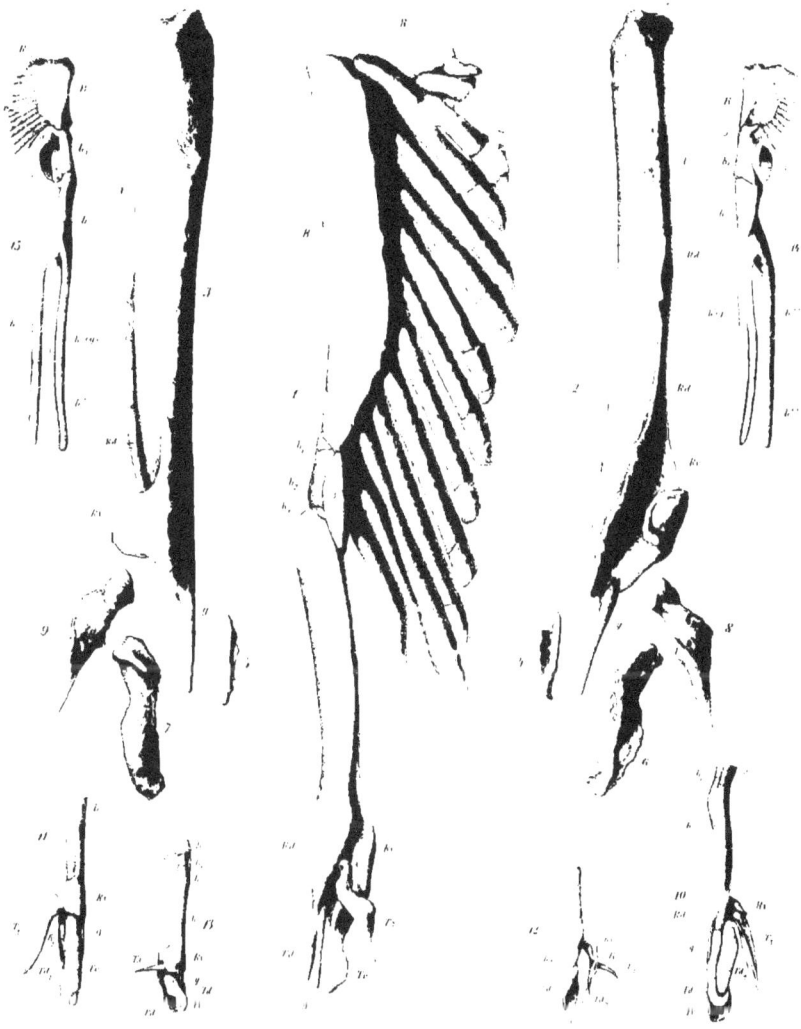

Fig. 1–9 Somniosus microcephalus   Fig. 10. 11 Acanthias vulgaris   Fig. 12. 13 Spinax niger
Fig. 14–15 Chimaera monstrosa

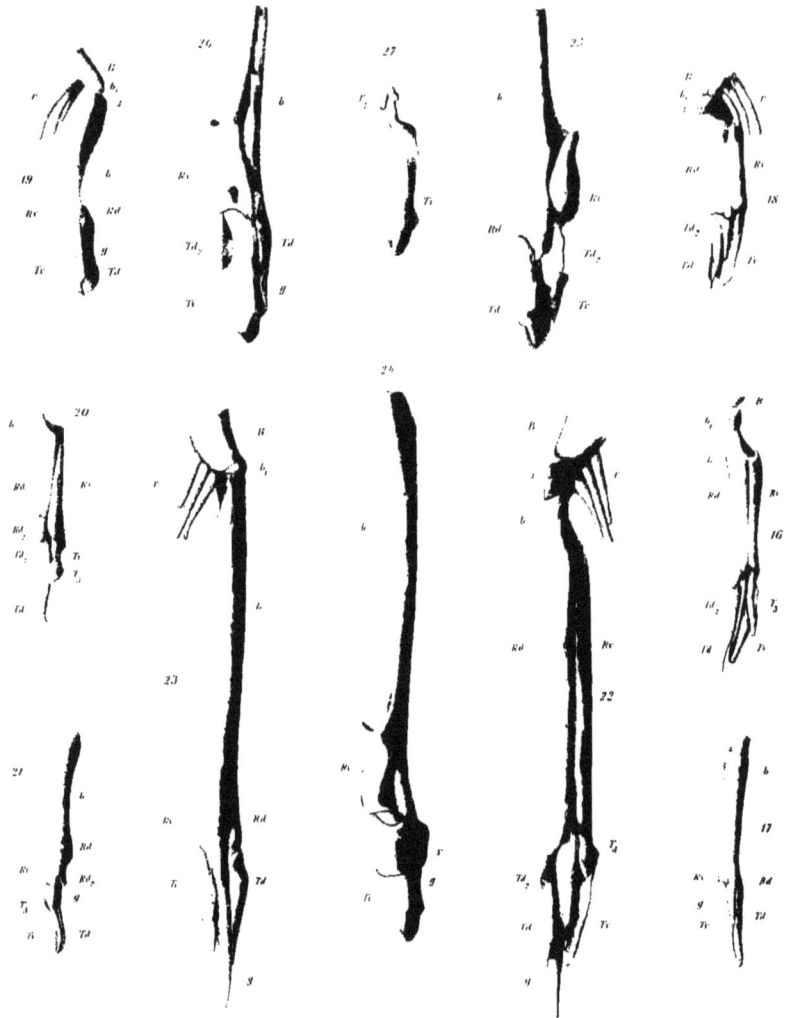

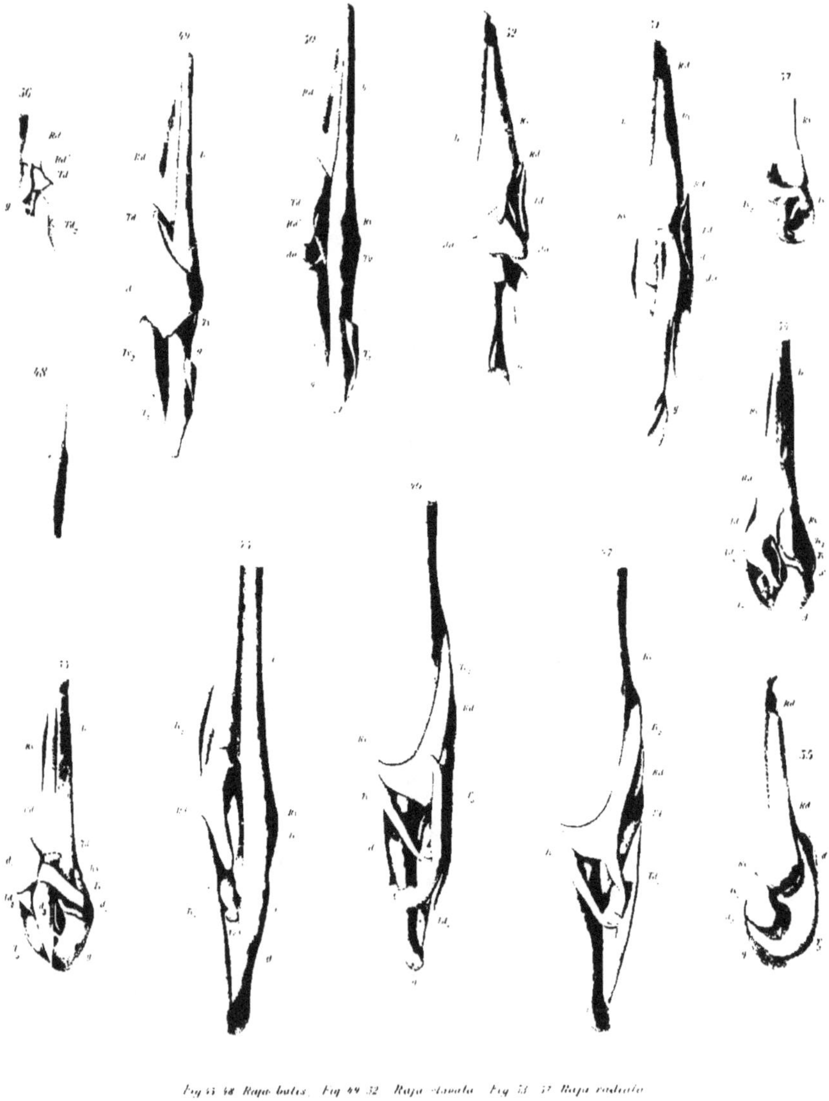

Fig 55-48 Raja batis. Fig 44-52 Raja lintea. Fig 53-57 Raja radiata

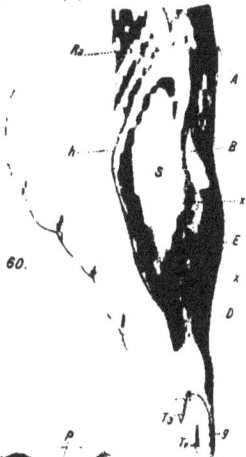

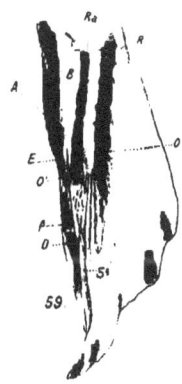

Fig. 58-62 Somniosus microcephalus, Fig. 63-64 Acanthias vulgaris ♀.

Hester Jungersen del.   Winkel & Magnussen impr.   Gantzen Schwist lithogr

Fig. 65-66: Scyllium stellare;   Fig. 67-68. Raja clavata;   Fig. 69-71: Chimæra monstrosa

www.ingramcontent.com/pod-product-compliance
Lightning Source LLC
Chambersburg PA
CBHW021948190326
41519CB00009B/1180